Stinking Philosophy!

Stinking Philosophy!

Smell Perception, Cognition, and Consciousness

Benjamin D. Young

The MIT Press
Cambridge, Massachusetts
London, England

The MIT Press would like to thank the anonymous peer reviewers who provided comments on drafts of this book. The generous work of academic experts is essential for establishing the authority and quality of our publications. We acknowledge with gratitude the contributions of these otherwise uncredited readers.

This book was set in Stone Serif and Stone Sans by Westchester Publishing Services. Printed and bound in the United States of America.

Library of Congress Cataloging-in-Publication Data

Names: Young, Benjamin D., author.
Title: Stinking philosophy! : smell perception, cognition, and consciousness / Benjamin Young.
Description: Smell perception, cognition, and conscio. | Cambridge, Massachusetts : The MIT Press, [2024] | Includes bibliographical references and index.
Identifiers: LCCN 2023042961 (print) | LCCN 2023042962 (ebook) | ISBN 9780262548885 (paperback) | ISBN 9780262379434 (epub) | ISBN 9780262379427 (pdf)
Subjects: LCSH: Smell—Philosophy.
Classification: LCC BD214 .Y68 2024 (print) | LCC BD214 (ebook) | DDC 128/.2—dc23/eng/20231205
LC record available at https://lccn.loc.gov/2023042961
LC ebook record available at https://lccn.loc.gov/2023042962

10 9 8 7 6 5 4 3 2 1

for
Jordana

Every man takes the limits of his own field of vision for the limits of the world.
—Arthur Schopenhauer (1891, p. 69)

Contents

An Apologetic Preface

You're sorry and I forgive you! But why did it take so long for philosophers to care about smell? Philosophy often progresses through its errors, but overlooking smell is a mistake that I hope to begin to rectify not by offering a comprehensive introduction to everything philosophers should know about smell and how its intricacies could have forestalled many quagmires of debates over the past century. The purpose of this book is not to shame philosophy for its stench. Rather, over the course of more than a decade, I have been arguing that studying smell provides a means of making lateral progress on a range of central debates in philosophy of mind and perception. *Stinking Philosophy* sets out to provide an unapologetic coverage of the philosophy of smell that, despite its meandering tone, is not intended for general readership. It is for philosophers interested in sensory qualities, smells, tastes, flavors, mental representation, nonconceptual content, concepts, and consciousness. The material assumes a background knowledge in all these debates, as well as a cursory knowledge of the chemosciences and philosophy of cognitive neuroscience.

With this purpose and audience in mind, what follows is a departure from previous manuscripts on olfactory philosophy such as Andreas Keller's *Philosophy of Olfactory Perception* (2017) and Ann-Sophie Barwich's *Smellosophy* (2020). Each provides a wonderful introduction to the vast amount of chemosensory knowledge required to understand smell and motivate the need for philosophers to focus on olfaction. The reader would do well to consult these books should my own introductions of experimental results relevant to the philosophical areas seem too brief for your needs. Additionally, for those looking for a recent collection on contemporary philosophical debates, *Theoretical Perspectives on Smell* (Keller & Young, 2023) serves as an excellent starting point for philosophers looking to gain a foothold in this rapidly developing research area. Each of these books differ markedly from what I proffer. *Stinking Philosophy* provides an overview of what looks like a range of disparate topics. What I set out to do is progressively elucidate different interesting questions about smell and, in doing so, weave together a decade of philosophical investigations into a coherent research project.

We, of course, begin by asking what smells are—a question that surprisingly and, despite the best efforts in philosophy and the chemosciences, is still rather vexing. Once I provide my own method for addressing this question as a nested set of issues, chapter 3 turns to using this methodology to demarcate orthonasal (from the front of the nose) smell from taste, which then brings up the question of the representation format of olfactory sensory transduction, perception, and cognition. Chapter 4 examines the format employed by the olfactory system in representing smells and argues that smell employs a nonconceptual format of representation. As the central chapter, it provides a novel understanding of nonconceptual content, an argument that functional noncompositional systems of representation are biologically plausible, and an interpretation of the discrepancy between our powerful ability to detect and discriminate smells yet meager capacity to identify odors by name. If I am right about the format of smell (and it is a big if), then conceiving of olfactory conceptual states presents a challenge, especially if these employ a dissimilar format from our linguistic conceptual repertoire. Thus, chapter 5 provides a progressive defense of how olfactory cognitive categorization preserves the alternative format of compositionality found in our perception of smell. Having covered the nature of smells, their perceptual representational format, and cognitive categorization, we then turn to our consciousness of smells. Chapter 6 shows how smell provides an empirically viable means of understanding the distinctions between different kinds of consciousness, such that it can be shown that most of our olfactory experiences occur in the absence of awareness yet always have a qualitative character. Chapter 7 provides closure in weaving together my research on olfactory philosophy by revisiting an old hobbyhorse. The chapter renews my claim that scientific theories of consciousness are inadequate or false based on their generalization from vision and neglect of smell. Thus, the book begins by showing how philosophy can benefit from studying smell and concludes by further demonstrating how neuroscience can also be enriched by attending to olfactory philosophy.

Yes, that is rather a lot to offer in a book this size. We have much to cover—if only you had been sorry sooner. My approach to all of these topics will be philosophical conceptual analysis and argumentation that is heavily empirically informed. Empirically informed olfactory philosophy is the default method for conducting research in this small area. Often, experimental results will be used as the evidential basis of my premises, although, at times, it might be certain camps of chemoscientific approaches that I will inhabit in developing my approach and resultant theory. Those unfamiliar with chemoscience and the olfactory system will be provided with a brief introduction in chapter 1. When this is inadequate for the reader's needs, I suggest getting uncomfortable with tracking down the cited scientific literature—this is, after all, an unapologetic treatment, and taxing you for neglecting smell seems like a fair trade-off for forgiveness.

Having been rebuked for neglecting smell and hopefully motivated to continue reading, I humbly apologize for the rhetorical flourishes above. The rest of the book will progress in the genteel fashion you have come to expect from philosophy books.

Acknowledgments

My research on smell began as a single chapter of my PhD dissertation concerning the tracking argument for nonconceptual content. Smelling quickly took over as a means of handling many of the issues I wanted to tackle but, from a lateral access point, bypassing many entrenched dogmatic starting points within philosophy. The book bears a resemblance to the structure of the dissertation but is in no way a replication thereof. I would like to thank my fellow graduate students, colleagues, and teachers from this period whose constructive criticism and comments are most appreciated: Jacob Berger, Ned Block, Richard Brown, David Chalmers, Jennifer Corns, Jerry Fodor, Lin Fou, Jim Hitt, Alexander Kiefer, Christopher Peacocke, Michal Klincewicz, Pete Mandik, Myrto Mylopoulos, David Pereplyotchik, and Josh Wiesberg.[1]

My advisors and mentors Bence Nanay, David Papineau, Jesse Prinz, David Rosenthal, Richard Samuels, Gabe Segal, and Richard Sorabji each contributed their own thinking to my theory development over the years. Some embodied the philosophical ideal of clarity and precision in stating and explaining philosophical arguments, some with their dedication to empirically informed philosophy, some for their deep engagement and broad knowledge of philosophy across time that still inspires me to try to situate my own views against these predecessors. But, above all, each spent innumerable hours discussing philosophy with me in such a joyous way—philosophy, after all, should be rigorously enjoyable. I have done my utmost to state my arguments simply and with precision while trying to keep this fun to read. Any success in this regard should be attributed to their philosophical mentoring, and of course, any failures are purely my own. I would also like to thank my postdoctoral supervisors Hilla Jacobson and Ran Hassin whose initial faith in my research on smell and willingness to fund it provided early career stability. Ran's dedication to disabusing me of just relying on neuroimaging results and cultivating a profound respect for behavioral methods deserves a special debt of gratitude that is apparent throughout the book.

Having great colleagues always makes doing philosophy so much easier, and my current colleagues at UNR are simply fantastic. I cannot count the number of work-in-progress seminars for which they patiently provided feedback on yet another paper on smell. In particular, I would like to acknowledge the central role Carlos Mariscal played in keeping this project going through kind words and peacefully weathering my constant diatribes about olfaction.

I would, of course, be remiss if I did not mention the olfactory philosophers who have been part of the journey developing this niche area. Thanks to Bill Lycan for pushing us all to stop slighting smell, as well as Solveig Aasen, Clare Batty, Ann-Sophie Barwich, Andreas Keller, Louise Richardson, and Barry Smith for their insightful questions, comments, and debates over the years both through publication and for their review work that always improved my papers. I would also like to thank all the olfactory philosophers who attended the two smell workshops I organized in 2021 and 2022, as well as all the contributors to the edited collection *Theoretical Perspectives on Smell*.

As an empirically informed philosopher, I would also like to thank my colleagues across the sciences who have been extremely helpful in increasing my knowledge of the chemosciences and who have been a tremendous resource over the years: Artin Arshamian, Thomas Hummel, Johan Lundstrom, Maria Larsson, Asifa Majid, Dennis Mathew, Jonas Oloffson, Noam Sobel, and Tali Weiss.

Lastly, I would like to thank my family. My parents deserve a huge thank you for all their help and support, especially for the hours of childcare that allowed me to steal away some afternoons to hectically write. My in-laws also deserve gratitude for their support over the years. My kids, Gabriel, Zachary, and Jonah, have all brought their unique intellectual gifts and curiosity to bear on my research and provided much-needed comedic interludes. For example, I still owe them a paper titled "Smelling Farts and Tasting Burps." Our dinner-table conversation often ends up about stinky matters, much to the chagrin of my wife, who wonderfully attempts to keep us civilized. Jordana, this book owes its existence to your loving support and patience with a smelly philosopher. Thank you. I love you not just because you smell nice.

1 Introduction

Smells are a curious sort. We don't pay olfaction much attention. Words often don't seem to capture odors' evanescent existence. Our conscious experiences seem to attribute little weight to smelling in the causal structures of our daily schedule. And yet, like many things neglected, invisible to the eye, and seemingly mysterious, once we begin examining their nature, we are presented with an abundance of riches. For more than a decade, I have been researching smell from an empirically informed philosophical perspective. The purpose of the book is thus to weave together the strands of my work on olfactory philosophy into a coherent research project covering what might be considered the main topics required to understand the nature of our experiences of smell.[1]

Research on smell in both the sciences and philosophy has only just begun to mature, having been neglected for a long period. As a research frontier, many of the debates and central questions are still being staked out, and the theoretical positions are still being demarcated. In some ways, as a new field, we are still attempting to grasp the central questions that need to be asked. While there are many fundamental issues that require elucidation about smell, I have focused on three that I think are the most central to generating an understanding of the nature of our experience of smells: what smells are, how smells are represented in our perceptual and cognitive states, and the nature of olfactory consciousness. There are many important philosophical issues not touched upon and left aside in what follows, such as the aesthetic nature of smell, the ethics of the flavor and fragrance industry, and the full range of topics that might be explored in relation to nonhuman olfaction. Additionally, there are even facets of the areas I will be covering that will be set aside, such as the role of scientific methods and techniques in generating new discoveries, the metaphysics of sensory qualities, and the relation between the different types and levels of processing relative to a given scientific domain and, of course, across domains. Doing justice to all these important issues and topics is beyond the scope of this book. Olfactory philosophy is, after all, a rather new area of research. There is much room for others to join us and explore the riches.

What will be addressed across the next six chapters are what I take to be the most pressing questions any adequate theory of smell mut be able to handle:

What are smells?

How do we represent smells within perception and cognition?

In what sense are we conscious of smells?

What I hope to make apparent as the book progresses is how my research over the years fits together to form a comprehensive view whereby my answers to each of these issues are all interrelated. The trinity of questions provides the basic structure of the book, with two chapters devoted to each question.

The first set of chapters are devoted to updating my framework for what might be considered smells, which is then extended to the other chemosenses and, in particular, taste. The middle chapters focus on how we represent smells within perceptual experience and olfactory cognition, with a particular focus on the format employed by the olfactory system and cognitive states concerning smell categories. The book then wraps up with two chapters devoted to olfactory consciousness, in particular how our smell perception requires that we rethink the taxonomic kinds of consciousness and their interrelations. The concluding chapter then argues that neuroscientific theories of consciousness would benefit from spending a little more time sniffing around chemosensory science and reading some olfactory philosophy. Thus, the book begins by showing how philosophical research is enriched within philosophy of mind and perception on such topics as sensory qualities, object perception, perceptual modality individuation, nonconceptual content, and kinds of consciousness and concludes by showing how olfactory philosophy is additionally of use in progressing scientific debates.

The next section provides a cursory overview of smell's importance throughout life that further motivates why, given its outsized role in shaping our existence, its neglect is all the more outstanding. Section 1.2

introduces the olfactory system, assuming feedforward propagation starting at the nostrils and progressing through cortical processing. The focus will be primarily on anatomical progression, with minor notes on functional roles to situate the further chapters that will provide greater treatments of the empirical research as needed. Section 1.3 summarizes the chapters in more depth, sketches out the book's progression, and notes what is completely new in the book, as well as what previous publications are being drawn from. These summaries should allow you to pick and choose your own smell adventure should my narrative become too cumbersome and taxing for your expertise. The chapter then concludes by noting those areas that will not be covered. There are many areas of olfactory philosophy that are still ripe for exploration, which were set aside in wrestling this book into existence.

1.1 Olfaction's Foundational Phylogenetic and Ontogenetic Status

A comprehensive philosophical treatment of our sense of smell is important because the olfactory system is phylogenetically and ontogenetically more basic than either vision or language, which are often employed as our default starting points for theorizing in philosophy. Ontogenetically, the olfactory system is employed even by fetuses and influences such important aspects of daily life as the identification of kin, food preferences, social selection, and our choice of mates. Phylogenetically, olfaction predates all other modalities, making inferences and extrapolations from animal models of olfaction to humans empirically more viable. Since the olfactory system develops before our visual or linguistic abilities and predates them, its study provides a rich access point for studying our minds while placing us on a continuum with other biological organisms.

Ontogenetically, olfaction develops well before the visual and auditory systems, allowing conclusions to be drawn about the content and consciousness of human minds well before the development of other perceptual systems. The olfactory system is fully functional during gestation, which is not the case for audition or vision. Although newborns can orient themselves to sounds from birth and show a preference for their mother's voice, it takes between one and two weeks for the auditory system to develop (or at the very least for the fluid within the cochlea to dissipate), which is necessary for them to encode sounds. Indeed, the length of time that it takes the visual system to mature fully is quite well documented, including the years it takes for children to develop visual object perception fully, including the psychological rules of feature integration and object constancy.

Infants do not immediately recognize their parents or caregivers by sight or sound, but they do identify them using olfaction from birth onward. Anosmic rats do not survive, since the sense of smell is necessary for infants to identify their own mother and orient themselves to her nipples. Young infants (mere minutes old) are able to recognize their mother's lactating nipples via the sense of smell and show preferential head movements in an attempt to orient themselves (R. H. Porter & Winberg, 1999; Schaal, 2012).[2] Additionally, young infants have been shown to be able to discriminate the smell of their mother's milk from others and to increase motor coordination crawling efficiency toward the odor (Hym et al., 2020).

Olfaction is a basic sense that shapes how we perceive and interact with our environment even before birth. The sense of smell in utero is responsible for our future food choices: what one's mother ingests while pregnant has a significant impact upon the tastes and smells of the objects that one is willing to consume for the rest of one's life. Olfaction is responsible for our food preferences (P. Rozin et al., 1986), allows us to recognize if food is fit for ingestion (Fallon & Rozin, 1983), and determines which new foods we are willing to try in adulthood based on flavor principles and ethnic culinary styles that are learned in childhood (E. Rozin 1983; P. Rozin, 1978).

Furthermore, olfaction is an important factor in our ability to determine kin relationship. Infants show a preference for their mother's breast pad, which can only be determined by smell (Russell, 1976). Mothers have the capacity to discriminate between the smell of their infant and the odor of other children (R. H. Porter et al., 1983). Aside from infant–mother kin detection, nuclear family members can also recognize the smell of their kin (R. H. Porter et al., 1986). Additionally, a very recent study demonstrated that we subconsciously smell ourselves and others, which mediates friendship selection based on shared underlying chemical similarities—that is, we like people who smell like us (Raverby et al., 2022), which nicely foreshadows my arguments in chapters 2 and 6 that the molecular features of the odorant partially determine olfactory quality (what it smells like), as well as the fact that these qualitative sensory states occur unconsciously and modulate human social acquaintance selection and behavior.

Phylogenetically, olfaction predates all other modalities. Olfaction's basic nature at times causes interesting conundrums. In humans, the olfactory bulb (OB) projects directly to the cortex, bypassing the thalamus, thus making it the only modality that necessarily requires the cortex for conscious awareness of stimuli. Yet, olfaction is the most basic form of sensing the environment, having evolved in most species well before the rise of the cortex.

A further implication of olfaction's ontogenetic and phylogenetic primitive nature is its validity in deriving conclusions about human behavior or psychology from animal models. Mammalian olfaction is highly conserved (Ache & Young, 2005), such that similar structures and functional processing occur across species. Moreover, the human olfactory system is not very different from that of goats and guinea pigs (Sela & Sobel, 2010). But the place of olfaction in human behavior is largely diminished in comparison to most mammals (Stevenson, 2009b). Humans mistakenly downplay and distrust their olfactory acuity. We might not think we are as gifted olfactory perceivers as bloodhounds or rats, but our olfactory capacities are in fact excellent.

Animal models of human behavior and cognition are an important research tool in biology and psychology. Olfaction is shared across species, thereby providing a tremendous body of research that can be validly used to make inferences from studies on other creatures about our olfactory states. The similar structure and functional organization of the olfactory system across species allows all sorts of inferences about how humans might encode scents.

Olfaction is responsible for our ability to navigate our environment, fulfilling our biological needs and shaping social interactions. Additionally, olfaction guides our choice of mates, our ability to identify our family, and our ascertaining parentage of one's offspring (S. C. Roberts et al., 2020). Lastly, smell is responsible for our selection of social circles.[3] Whether due to bodily odors, diet, or general environmental odorants embedded in their clothes, we show an implicit social preference towards those that smell more similar to us. Whether we are aware of it or not, our sense of smell shapes major aspects of our lives.

1.2 Introducing the Olfactory System

Olfaction is different from the other perceptual modalities in ways that have serious import for the study of perception, cognition, and consciousness, as well as the cultural mediation of each of these (Gross, 2019). As an introduction to the olfactory system's unique architecture and stimulus processing, this section proceeds by discussing the olfactory system's anatomy and its functional organization. What is offered is a cursory introduction to the olfactory system that should enable those without a background in the chemosciences not to get lost in the major discussions of anatomical structures and neural processing discussed throughout the book.

The focus of this book is solely on smell, aside from chapter 3 on taste and flavor. There will be little coverage of the other chemosenses (taste and the flavor system) or the other chemosensory subsystems within the nose such as the vomeronasal system (Dulac, 2000; Halpern, 1987; Keverne, 1999; Meredith, 1991), which is implicated in pheromone encoding in nonhumans, or the trigeminal system (Doty, 1995; Hummel, 2000; Hummel & Livermore, 2002), which is responsible for the sensation of irritation and pain responses associated with olfactory stimuli. Aside from issues of scope, I have not considered these last two mammalian chemosensory subsystems because their existence (i.e., vomeronasal and pheromones; Doty, 2010) and utilization by humans is unclear (Meredith, 1991; Sela & Sobel, 2010; Witt & Hummel, 2006).

Before introducing the olfactory system, it is worth noting the range of molecules that the human olfactory system transduces. We can detect the smell of a volatile molecular species (molecular weight<294 daltons) with surface activity, low polarity, water solubility, high vapor pressure, and high lipophilicity (Ohloff, 1986). However, in order for an odorant to be smelled, it must first reach the olfactory epithelium by traversing the nasal passage and the mucosa layer within our nostrils.

Odorants reach the olfactory epithelium and olfactory receptor neurons (ORNs) orthonasally or retronasally. Orthonasally (from the front of the nose via the nostrils), the odorant reaches the epithelium either through diffusion from high levels of concentration to lower levels or through actively sniffing the odorant. Alternatively, an odorant might arrive from the back of the throat via retronasal olfaction (Hornung & Enns, 1986). The retronasal system is highly involved in flavor and gustation and has its own functional organization, as well as rules of odorant transduction. While retronasal olfaction is fascinating, it will be discussed and bracketed in chapter 3, since the book is primarily focused on orthonasal olfaction, and all arguments and evidence offered should be understood as only relative to what I will consider to be smell in this narrower sense.

1.2.1 The Nostrils: Shape and Number Matter

Ascending through the anatomy of the olfactory system, an odorant enters the nose through the nostrils or via the back of the throat and becomes encased in the mucus surrounding the olfactory epithelium. It would be easy to skip ahead to the nature of receptor cells and transduction processes within the primary sensory area of the olfactory epithelium. Yet, this would grossly overlook the importance of the nostrils and mucus layer. Both the shape of the nostrils and their number are important. Having two nostrils allows us to determine the location of a smell in a similar fashion to sounds in audition using two ears. There are small differences in time and intensity between a sound arriving at one ear versus the other, as well as between a smell arriving at one nostril versus the other (von Bekesy, 1964). Additionally, the shape of the nostrils is of importance because they are asymmetrical in airflow, which switches every couple of hours (Bojsen-Moller & Fahrenkrug, 1971), changing each nostril's sensitivity to odorants depending upon the rate of airflow at different sorption rates (Sobel et al., 1999a).

In general, the rate of airflow is higher in one nostril than the other, which is caused by swelling of the nasal turbine that increases airflow resistance in one nostril as opposed to the other (Bojsen-Moller & Fahrenkrug, 1971; Hasegawa & Kern, 1977; Principato & Ozenberger, 1970). The rate of increased airflow also alternates between nostrils in accordance with an ultradian rhythm (Gilbert & Rosenwasser, 1987; Mirza et al., 1997). When the information regarding airflow compared by nostril is combined with the sorbency rates across nostrils in accordance with information about sniffing, this generates the result that each nostril is tuned to odorants that sorb to the mucus at the current flow rate in that nostril (DePay & Hornung, 2002; Sobel et al., 1999a). Each nostril conveys a different olfactory percept to the brain, which depends upon airflow and sorbency rates. That each nostril creates a different olfactory percept is substantiated by W. Zhou and Chen (2009), who demonstrate that binaural rivalry exists between the nostrils. Their research shows that perceptual rivalry can occur within the olfactory system, such that "alternating odor percepts [occur] when two different odorants are presented to the two nostrils" (p. 1564).

Moreover, it seems that we unconsciously smell in stereo, and this is utilized in odor navigation (Y. Wu et al., 2020); that perceived intensity of an odorant presented to only one nostril is inversely related to the degree of perceived airflow in the contralateral side (Yao et al., 2020); and that odorant identification seems to be better with the left nostril (Zang et al., 2020). Thus, we utilize both nostrils in a myriad of ways that subserve a number of upstream olfactory functions.

1.2.2 Mucus Matters

Mucus plays an invaluable role within the olfactory system. Mucus coats the olfactory epithelium and is produced by the cells of Bowman's gland. Mucus is responsible for immune function, enzymatic conversion of odorants, stimuli transduction, and removal of olfactory stimuli (Robert-Hazotte et al., 2019a).[4] Additionally, it contains odorant-binding proteins for transferring the odorants through the mucus layer (Pelosi, 2001; Pevsner et al., 1985), which might determine if a molecule is odorous based on its transport features and which might be further used to determine the parameters of olfactory dimension of perceivable olfactory qualities (Mayhew et al., 2022).

The sorbency of odorants for traversing the mucus to reach the cilia of ORNs plays a vital role in how the olfactory system computes olfactory stimuli. The mucus layer in combination with the sorbency of odorants allows the olfactory system to process peripheral environmental shifts continually. However, this requires slow temporal processing speeds. Additionally, it might enable the olfactory system to account for changes in concentration levels across a given type of stimuli. Lastly, mucus and sorbency rates are needed to calculate the olfactory stimuli. Especially with regard to sniffing odorants, the sorbency of chemical stimuli and rate of airflow needs to be accounted for in determining the olfactory quality (what the odorant smells like) of an olfactory stimulus. High airflows will optimize perception of higher sorption rate odorants, and low airflows will optimize perception of lower sorption rate odorants, both of which are required to calculate the olfactory stimulus (Keyhani et al., 1997; Mozell et al., 1991). More recently, it has been shown that, at least with food odorants, the metabolites within olfactory mucus might also play an active role in changing the odor profile (Ijichi et al., 2019), as well as increasing the odor detection threshold for some odorants presented subliminally (Robert-Hazotte et al., 2019b).

1.2.3 The Olfactory Epithelium and Olfactory Receptor Neurons

Having traversed the nasal cavity and mucus, an odorant must come into contact with ORNs within the olfactory epithelium to be encoded. The olfactory epithelium consists of a sheet of receptor neurons (and basal cells poised to become ORNs or support cells) composed of roughly one thousand different types of receptor cells as determined genetically according to their ability to produce proteins to which the odorant molecules bind (Buck & Axel, 1991). In humans, this number decreases to between three hundred and four hundred types of ORNs (but just think of the combinatorial explosion for transduction when compared with the very limited three to six receptor types in vision). While roughly 3 percent of the mammalian genome is devoted to olfactory receptor formation, the vast majority of these genes are pseudogenes (roughly two-thirds), which do not generate ORNs (Rouquier et al., 2000). The number of pseudogenes is similar across mammals, although a decrease in pseudogenes occurs in species with less acute color perception.

ORNs are unique in two important ways. First, they regenerate over time, with a life cycle spanning a month to a year (Hinds et al., 1984; Mackay-Sim & Kittel, 1991; for a state-of-the-art review, see Tufo et al., 2022). The function of ORN neurogenesis is debatable, but the most traditional explanation is that regeneration is required to repair the damage caused to ORNs by being exposed to the environment. However, receptor regeneration over time could play a functional role in stimuli transduction by allowing the olfactory system further plasticity for sensitivity to novel stimuli as the receptors become more prevalent within a constantly evolving environment (Cummings & Belluscio, 2008),[5] as well as developing and maintaining a chemospatial map for odorant identity for ORNs as they project to the OB (Dorrego-Rivas & Grubb, 2022). Second, ORNs are special because they are exposed to the external environment, thereby coming into direct contact with olfactory stimuli. An ORN sends its dendrite into the mucosal layer, terminating in the olfactory knob that contains between three and fifty nonmotile olfactory cilia (Morrison & Costanza, 1990, 1992).

Both epithelia contain roughly six million ORNs, each of which is capable of binding to many different types of odorants (at a given time) based on the protein structures of their receptor sites. Thus, a typical odorant activates many different classes of receptor neurons but to different degrees. The olfactory system's functional organization for stimuli encoding further differentiates it from vision and audition. ORNs do not generate a sensory typology of chemotopic maps at the receptor level analogous to retinotopic maps and columnar organization within the visual system and sound-wave transduction by the cochlea within the auditory system. A chemotopic map is also not available at the OB's glomeruli and mitral cells. While ORNs are selectively sensitive to different odors to different degrees, there is no clear chemotopic mapping within the olfactory epithelium, since each type of ORN is diffused in a random manner throughout the olfactory epithelium (Yaksi et al., 2009). Chemotopic organization might occur with the glomeruli that received input from only one type of ORN from across the olfactory epithelium.

1.2.4 The Olfactory Bulbs

The axons of the ORN project through the bone of the nasal cavity to the glomeruli in the olfactory bulb. Just as the olfactory system operates using two distinct nostrils and epithelia, we have two distinct OBs. Each OB consists of six layers arranged in concentric rings similar to the structure of an onion (Greer et al., 1981; Shepherd, 1972) in the following order: (1) the olfactory nerve, (2) the glomerular layer, (3) the external plexiform layer, (4) the mitral cell layer, (5) the internal plexiform layer, and (6) the granule cell layer (Kratskin & Belluzzi, 2003). For the purposes of this book, I shall only discuss the glomerular and mitral layers and leave aside the other parts of the OB.

Glomeruli are spherical clusters of axons from the ORNs, dendrites from the mitral cells, and dendrites from local interneurons that modulate the connections between sensory neuron axon terminals and mitral cell dendrites. Aside from the spherically shaped glomeruli, the OB also contains mitral cells whose role is to act as excitatory connections with their axons traveling to the olfactory cortex (OC).

Each glomerulus's selective sensitivity to a particular odorant might form the beginning of a chemotopic map. However, they also have a different base firing rate for a secondary set of odorants involving mitral cells. Furthermore, as will be argued in greater depth in chapter 4, the OB encodes odorants by implementing a functionally compositional system of representation that does not obey classical concatenative compositionality. Within the OB, there is no strict odorant encoding and certainly no isomorphic mapping of odorant to receptor types as is the case with the cochlea and the cones within the retina. Rather, each stimulus is encoded piecemeal by the activation patterns of the glomeruli throughout regions of the OB. The distributed encoding of olfactory stimuli, in combination with neurogenesis, allows for a theoretical capacity of smelling a nearly infinite number of odorants.

The anatomical structures within the OB play a more robust functional role in stimuli encoding than the receptor sites in other sensory modalities, making comparisons between the structural hierarchy of the other perceptual systems and olfaction difficult. The OB is not analogous to the rods and cones or ganglion within the retina, subcortical relays such as the lateral geniculate nucleus (LGN) in the thalamus, or the cortex such as the primary visual cortex (V1) in the occipital cortex. The OB performs a greater computational role than the first two options, and even though it is considered to be cortical (based on its location within the skull and proximity to the cortex), the OB is not thought to be as computationally sophisticated as the cortex (although chapter 7 will discuss the functional role of the OB in greater depth).

1.2.5 The Olfactory Tract

Moving upward in the olfactory system, axons from the OB project via the olfactory tract to the primary OC (for a state-of-the-art study of human olfactory tract projections using diffusion magnetic resonance imaging [MRI], see Echevarria-Cooper et al., 2022). The olfactory system's anatomy is unique in not requiring thalamic connections before projecting to the cortex. There are no direct thalamic connections between the olfactory receptor sites in the olfactory epithelium and OB. Rather, the olfactory tract projects directly to the cortex.

The olfactory tract runs ipsilaterally from each OB directly to the primary OC. The lack of contralateral projection or any form of crossover makes olfaction unique, since information from the right receptors goes to the right cortex and left receptors to the left cortex (unlike the other sensory modalities that integrate incoming stimuli before projecting to the cortex), such that lateralized cortical processing centers are implicated in processing different aspects of olfactory perception (Cavelius et al., 2022). The olfactory tract's ipsilateral projections and lack of contralateral connections are still the traditional organization of olfactory anatomy. Yet, some have begun to question this dogma using functional MRI (fMRI) findings that are suggestive of the existence of contralateral pathways (Cross et al., 2006; McBride & Slotnick, 1997; J. Porter et al., 2005; Savic et al., 2000; Uva & de Curtis, 2004; D. A. Wilson, 1997). Moreover, recent experimental work has shown lateral interconnections in mice between the glomeruli of OBs mediated by their mitral and tufted cells (Grobman et al., 2018), which might partially explain how, despite having ipsilateral cortical projects without crossovers and functional localization further upstream, we nonetheless subjectively report having experiences of unified olfactory percepts (Dalal et al., 2020).

1.2.6 Primary Olfactory Cortex, Piriform Cortex, and Thalamic Relays

The OB connects via the olfactory tract directly to the OC.[6] The primary OC consists of all brain regions that receive direct input from the mitral and tufted cell axons of the OB (Allison, 1954; Carmichael et al., 1994; de Olmos et al., 1978; Haberly, 2001; J. L. Price, 1973, 1987, 1990; Shipley, 1995). The primary OC includes the entorhinal cortex, periamygdaloid cortex, cortical amygdaloid nucleus, piriform cortex (PC), olfactory tubercle (OT), tenia tecta, and the anterior olfactory nucleus. The entorhinal cortex projects to the hippocampus, while the PC projects to the orbitofrontal cortex (OFC), the insula, and the mediodorsal nucleus of the thalamus (MDNT), which also gains incoming stimulation from the OT (Mackay-Sim & Royet, 2006). The PC accounts for the largest portion of the primary OC and lies at the junction of the temporal and frontal lobes, as well as fusing into the anterior cortical nucleus of the amygdala.[7] However, given the size and range of these cortical areas, the definition of what constitutes the primary OC is far from functional, and there have been some who suggest abandoning the current definition (Haberly, 2001; Sobel et al., 2003).

The primary OC also encodes airflow to allow for olfactory constancy (Teghtsoonian et al., 1978, 1982; Teghtsoonian & Teghtsoonian, 1984). Given that odorants are unevenly dispersed throughout the environment and airflow varies between the nostrils, cortical monitoring allows the system to account for these fluctuations within the olfactory object by comparing the incoming stimuli against airflow and sniffing. For example, if a high concentration of odorant with a high velocity is presented to one nostril and a low dose with a low velocity is presented to the other, then this might be one of the mechanisms allowing the system to surmise that it is the same olfactory entity.

Aside from encoding nasal airflow and sniff rate, the OC has areas devoted to the categorization of odorants as well as their identification (see chapter 4 for further discussion). Thus, odorant coding would seem to come to a head within the cortex. However, each layer plays a role in olfactory encoding. Since we can discriminate more odorants than we have receptor types, olfactory odor coding cannot occur at the ORNs, thereby ruling out a one-to-one receptor-to-odorant mapping scheme. Current research suggests that odorant encoding is a combination of the activity summed across the ORNs and glomeruli in a manner that is both spatially and temporally distributed. Odorant coding is a combination of temporal and spatial encodings for a complete percept. Compared to vision and audition, olfactory temporal processing is slow at about 150 milliseconds, given the need to traverse the mucosa layer (Firestein & Werblin, 1989), which already begins to suggest that olfactory perception, cognition, and consciousness might present differently than vision and that our everyday intuitions of how things appear based on retrospective introspective reports might not fit olfactory experience.

1.3 Chapter Summaries

The book weaves together my work from the past decade into a coherent coverage of research on a wide range of topics. The coverage is not just a historical unpacking. Rather, each chapter has been crafted to display the development of thinking and research, especially with regard to how my molecular structure theory (MST) of smell perception, theory of formative nonconceptual content (FNCC), and work on qualitative consciousness in the absence of conscious awareness all fit together. Those familiar with these theories may wish to skip to the completely new material in chapters 3, 5, 6, and 7. For those unfamiliar with olfactory philosophy, what follows is a brief summary of each chapter to elucidate the book's progression and to allow you to figure out what sparks your curiosity, although it should be noted that the progression across chapters assumes knowledge of the proceeding arguments and empirical evidence.

Chapter 2: What Are Smells?

The centrality of the question "What are smells?" cannot be escaped when beginning any exploration of how we perceive, represent, and are conscious of smells. Thus, it serves as the starting point. The study of smell is still rather nascent, making the initial question of the chapter pertinently tractable, as we still don't have a clear answer to this question. The seeming simplicity of the question hides that it is a rather complex set of issues requiring a multifaceted set of answers. Previously, I have argued that a comprehensive theory of smell must answer the following three nested questions:

1. What accounts for the olfactory quality of a smell?
2. What are the odorous objects represented within our smell experiences?
3. What are the distal entities that we perceive as smells?

The chapter summarizes what I think are the best answers to these questions as the initial starting point for exploring what smell has to offer philosophical research. The chapter concludes that smells are complex perceptual objects that not only smell but also have perceptible properties, including their concentration, intensity, and valence.

While the chapter is new in terms of being the most current version of MST, it is developed from a host of older papers, including "Smelling Matter" (2016), "Smelling Molecular Structure" (2019a), "The Many Problems of Distal Olfactory Perception" (2019b), "Perceiving Smellscapes" (2020), and "Maybe We Don't Smell Molecular Structure" (2023).

Chapter 3: Tasting Flavors

Chapter 3 adapts the explanatory framework from chapter 2 and employs the nested questions of what the object of perception is, how these are represented within experience, and what the distal entity we perceive is, in considering what our sense of taste is. I further argue that we can use this framework to account for other chemosenses, such that we taste consumable items placed within the oral cavity with their flavorful sensory qualities that generate a non-decomposable format of flavor experience. Often, what we consider to be the taste of food and drink is not just its gustatory qualities but rather a complex multisensory experience that includes retronasal olfaction, gustation, somatosensation (including thermal and nociception), and chemothesis. Each of these separate sensory channels provides access to the multifaceted object of flavor perception, which compose a unified perceptual experience of flavor. The central thesis of this chapter is that flavor perception is accomplished by a host of sensory systems that combine to yield the unimodal perceptual modality of taste. To establish the claim that we taste flavors, the paper adopts a pluralist conception of the senses and shows that we can employ the same explanatory strategy for the chemosenses of smell and taste.

"Tasting Flavors" is new material that is a sister project of "Smelling Odors and Tasting Flavors" (2023). I began researching both while on a surf trip in Big Sur in 2018 and had originally planned this as a single chapter, but it became so long and disjointed that the project was split into two separate pieces based on their distinct conclusions.

Chapter 4: Formatting Odors

Chapter 4 focuses upon the representational format of olfactory perception. To do so, it first establishes that our experiences of smells are representational. Then, it transitions to accounting for the kind of representational system implemented by smell. Once it is allowed that olfactory states are representational, the nature of their representational format is identified in terms of the type of compositionality implemented within the olfactory system. Analyzing the representational format employed by the olfactory system has the added value of yielding an understanding of nonconceptual content as nonconcatenative compositionally formatted mental states. The chapter shows how olfaction provides a novel means of reinvigorating the debate about nonconceptual content, providing an empirically viable example of how a neurobiological system could implement a nonconceptual representational systems, while also explaining the puzzling discrepancy between our gifted ability to detect and discriminate smells and our meager capacity to identify smells.

Chapter 4 further develops material from "Formative Non-Conceptual Content" (2015), "Smell's Puzzling Discrepancy" (2019d), and "Olfactory Imagery" (2019c), as well as updating material from my PhD dissertation. While a good deal of the evidence and arguments remain the same, the progression of material and structure of presentation nicely display how the theory that the olfactory system implements a nonclassical format of composition has developed and how it can accommodate experimental findings from across the different stages of olfactory processing, as well as how the format is conserved from sensory transduction through perceptual experiences.

Chapter 5: Pondering Smells

Chapter 5 addresses the question of how we think about and communicate about smells as categorical entities. What is quite clear from the outset is that if our cognitive states about smells implement the same format as that of the sensory and perceptual states from chapter 4, then olfaction will not employ concepts in anything like the classical philosophical notion of propositions with necessary and sufficient conditions. With this tension in mind, the chapter explores how it is that we can characterize smells across experiences, such that we can categorically think about smells across multiple instances and interpersonally communicate about smells.

Pondering smells certainly encompasses more than just their olfactory quality. Our cognitive states about smells also encompass their intensity, valence, hedonics, and associations that might be mediated by cultural practices, mode of living, and functional usage within daily life. Thus, the first thing to consider is whether humans' poor ability to name odors (covered in the last chapter) is universal or an artifact of linguistic and cultural practice that is predominantly measured within English speakers. The chapter reviews literature that suggests the difficulty in naming odors is meditated by language and culture, which then sets up the question of what our linguistic practices can tell us about the representational format of smell categories. It will be argued that the relation between our linguistic conceptual resources and olfactory representational states is symmetrical poor. We are bad at naming perceived odors, and verbal tags do not generally elicit olfactory representations of smells. The poor connection between language and olfaction provides further reason to think that if representing smells requires a complex system that is not a simple linear mapping relation of odorant to odor, then how we think about, ponder, and categorize smells will also follow a nonlinear pattern with both a complex quality space of smell qualities and combinatorial coding for our experience of the complex structural nature of smells and our categorization thereof. With the poor relation between language and olfaction in mind, the penultimate section explores how language and smell expertise might enhance our olfactory categorical abilities. The chapter's general conclusion is that the format of smell is preserved across sensory processing, perceptual, and cognitive states.

Aside from a small portion of material from "Smell's Puzzling Discrepancy" (2019d) concerning cross-cultural variance in odor identification abilities, the content and arguments within the chapter are completely new and not published elsewhere.

Chapter 6: Unconsciously Smelling

We may not always notice the world of odors enveloping us. Yet, they provide a powerful source of experience. Chapter 6 documents how olfactory consciousness provides a novel means of demarcating different kinds of consciousness. The chapter adapts the distinction between access and phenomenal consciousness in an empirically tractable fashion and provides a contrastive understanding of the relationship between conscious awareness and phenomenal conscious when understood as the qualitative character of an experience. It will be argued that we need to rethink the relationship between access consciousness and phenomenal consciousness, such that all olfactory states have a qualitative character that influences our behavior even when we are not subjectively aware of the smell—smells are always experienced qualitatively, even when the subject in unconsciously perceiving the smell. The chapter updates and redevelops published material from "Smelling Phenomenal" (2014) and "Quality-Space Theory in Olfaction" (2014) coauthored with Andreas Keller and David Rosenthal.

Chapter 7: Stinking Theories of Consciousness

Chapter 7 continues to show the importance of studying smell in generating an understanding of consciousness with a shift in focus to neurosciences' neglect of olfaction. Our sense of smell, the anatomical structure of the olfactory system, and its functional organization have profound consequences for the study of consciousness. While I have previously argued that the major scientific theories of consciousness are either false or inadequate as general theories because of their visuocentric methods and neglect of olfaction (Young, 2012), little has changed in a decade. The disproportionate dominance of vision is often simply taken for granted without even noting that experimental results must be relativized to visual consciousness. And while some are careful to note that they are developing a theory of visual consciousness, this is a rarity. For example, the entire debate couched in terms of the question "Is consciousness in the front or back of the brain?" (Boly et al., 2017; Koch et al., 2016; Odegaard et al., 2017; Storm et al., 2017) only makes sense if vision is assumed to be the default modality that universally generalizes. The mere starting tacit assumption of the debate will exclude the nuances of olfactory processing regardless of the kind of consciousness under consideration. The focus of this chapter is thus to review how neglecting olfaction has negatively impacted neuroscientific theories of consciousness (including empirically tractable philosophical theories), such that large portions of these theories are *still* either false or inadequate as general theories of consciousness. The chapter updates and redevelops never-before-published material from my PhD dissertation and builds upon research from my first publication "Stinking Consciousness" (2012). The chapter thus brings closure to the book that starts by showing how olfactory philosophy enhances debates within philosophy of mind and perception and concludes by showing the interdisciplinary import of philosophy of smell.

1.4 What Won't Be Covered

The rapid pace and emerging nature of research on smell requires that there are topics and findings that, by the time this book goes to press, I will not be able to cover, while others have been pragmatically omitted. There are a number of well-developed areas withing the chemosciences that deserve philosophical attention, but all I can offer for now is a promissory note.

Olfaction is claimed to be intimately connected to memory and emotions in a fashion that has led many to argue that the primary function of olfaction is not to perceive objects within scenes but rather to track chemical stimuli that are of ecological import to us purely in terms of their hedonics (Barwich, 2020; Keller, 2017). And some have claimed that olfaction is designed to make us aware of changes within our environment (Köster et al, 2014). While I do not share these theories' claims about the function of smell, they should be noted, since my central argument throughout the book—that we smell and represent smells as mereologically complex changing and repeatable entities within a smellscape—runs contrary to these. However, I hope it will become apparent in what follows that my view provides at least as coherent an alternative to these views.

Often, it is claimed that smell has unmediated connections to the limbic centers, which some identify as the neural hub for emotions, and thus olfaction is more emotional than the other senses. While the anatomical evidence seems viable, the further inference is certainly worthy of further philosophical scrutiny in terms of both the evidence provided for the inference and what exactly is being claimed about the relation to emotions. For instance, what theory of emotions is at play here? Given that the emotions are a rich area of philosophical investigation with many theories, some of which conceive of emotions less in terms of valence and more in terms of doxastic judgments using propositions or as mediate by our embodied states as filter through culturally embedded dispositions, it isn't clear why we are supposed to think the limbic connection on its own would be sufficient to make such strong claims. In general, the relation between smell and emotion is a fecund topic that deserves rigorous theoretical development.

How we remember smells is a vast and well-developed area within olfactory memory research, with experimental findings across a range of topics such as odor encoding, a smell's effect on retrieval, episodic memory, the anatomical and cortical basis of olfactory memory, and autobiographical memories. To the best of my knowledge, none of these areas have been explored in depth within philosophy, aside from a quick mention of the autobiographic odor memory bump in Young (2019c).

Further topics not covered are the aesthetics of smell, human chemosensory signaling (mentioned earlier in section 1.1) and its implication for 4E accounts of perception/cognition, as well as applied ethical topics relevant to the industrial production of odors and our personal use thereof.[8] And, of course, there must be a myriad of other areas fit for olfactory philosophical work that I am omitting simply due to a failure of imagination. May I extend these topics as an invitation to anyone specializing in these areas of philosophical research to sniff around and find out.

2 What Are Smells?

We are all overly familiar with the initial setup of philosophical discourses starting with a "What is *x*?" question, and in this instance, the philosophy of smell does not disappoint. The initial familiar setup of attempting to clarify the nature of smells conceptually serves as the focus of this chapter. The centrality of the question "What are smells?" cannot be escaped when beginning any exploration of how we perceive, represent, and are conscious of smells. And while such an opening chapter might seem derivative given Keller's (2017) and Barwich's (2020) similar starting points and our agreement that a simple answer is not forthcoming, I differ in thinking that the question itself is not singular. We cannot understand what smells are without giving a comprehensive account of a set of issues, all of which determine the nature of smells.[1]

The study of smell is still rather nascent, making the initial question of the chapter pertinently tractable. Yet, as with many aspects of olfactory research, we often find ourselves stuck in the middle because so many areas are still at an exploratory stage. Previously, I have argued (Young, 2016, 2019a, 2023) that a comprehensive theory of smell must answer the following three nested[2] questions[3]:

1. What accounts for the olfactory quality of a smell?
2. What are the odorous objects represented within our olfactory experiences?[4]
3. What are the distal entities that we perceive as smells?

The theory I have been developing addresses each of these questions by adverting to the molecular structure of chemical compounds composing an odor plume within a background array of a smellscape (Young, 2016, 2019a, 2019b, 2020).[5] Earlier versions of my theory were developed to account for how the olfactory quality of an odor is determined and employed to generate odor identity. In progressing this account, the chapter not only attempts to handle question 1 in relation to accounting for the determinates of the olfactory quality of complex odorant mixtures but also looks at how what we consider to be the object represented within smell experiences must be expanded beyond just the olfactory quality of smells. The seeming simplicity of the initial question "What are smells?" hides that it is a rather complex set of issues requiring a multifaceted set of answers. The chapter aims to give a summary of what I think are the best answers to questions 1–3 as the initial starting point for exploring what smell has to offer philosophical research, as well as building upon my previous treatments of the nature of odor objects. Smells are complex perceptual objects that not only smell but also have perceptible properties, including their concentration, intensity, and valence.

2.1 Question 1: What Accounts for the Olfactory Qualities of Smells?

We are immersed in a natural environment full of smells. Yet, our habituation to their persistence makes us prone to ignore this facet of our mundane existence. Trivially, we smell smells, whereby the most essential aspect of an odorant for generating odor identity is its perceived olfactory quality (what it smells like). The primary stimuli of the perceptual modality of smell are odorants—that is, diffused chemical compounds that traverse our nostrils, get stuck in the olfactory mucosa at the top of the nasal cavity, and eventually enervate our olfactory receptors. Yet, every aspect of this progression is still an open area of research. How odorants become odors is exactly what is under debate throughout the chemosciences and philosophy of smell.[6]

What I hope to make clear within this chapter is that smells are more than just the chemicals that cause them or even the olfactory qualities we ascribe to them as odors.[7] Olfactory quality might serve as the primary means of determining the identity of an odor, but considered as perceptual objects, smells have properties extending beyond their odor identity, including valence, intensity, and spatiotemporal properties. The identity criteria of odors with their set olfactory qualities should not be considered as co-extensive with the identity criteria of smells—the olfactory system allows us to perceive and track properties of the perceptible object of smell beyond olfactory qualities. But before we can move on to an analysis of what the objects of our perceptual experiences of odorific natural plumes are, we first need some means of determining the olfactory quality of a smell. What follows are reasons in favor of MST's claim that, for single odorants, the determinate of olfactory quality is the molecular structure of the odorants within a plume. How this then scales up to more complex naturalistic olfactory stimuli such as odor mixtures, odor objects (i.e., mereologically complex smells identifiable across token instances of variance in their composition, concentration, intensity, and valence), and smellscapes will be the focus of the sections that follow.

2.1.1 The Olfactory Quality of Single Odorants (Question 1: Beginnings)

Our philosophical starting point is often to begin with what is presented to us phenomenologically within experiences as given and then to reify this through conceptual analysis. Yet, our everyday olfactory experience and ordinary linguistic practices of referring to smells generate methodological issues at the outset of theorizing about the olfactory quality of smells. Our everyday practices, it has been argued, are inaccurate in determining the nature of smells (Young, 2016), our conscious experiences often incorrectly identify or fail to identify smells (B. C. Smith, 2023), and our overall linguistic competence in identifying smells by name is extremely limited (Young, 2019d). Thus, even the initial question of what method should be used in determining the olfactory quality of odorants is open for debate.

There are several possible measures of identifying the olfactory quality of a single type of odorant. Naming odors in terms of lexical tags might seem like the most intuitive means of determining identity conditions, but as noted in Young (2019d), humans are woefully bad across developmental stages at identifying smells by name, which might be explained by the olfactory system lacking connectivity with linguistic processing centers in the brain, as well as implementing an incompatible compositional format from our semantic conceptual repertoire mediated by vision (see chapter 4). But even if we relax the means of identification from matching exacting lexical tags to referents and instead employ the conceptual categorization of odors based on a typicality relation among members of some given type, we will face difficulties both in specifying the determinate of the typicality relations (Young, 2016) and in determining typehood of odor categories (Gilbert, 2008; see chapter 5). Using a measure of subjective recognition of an odorant's odorific quality will face similar issues, since recognizing a token odorant as being a given type of odor will require knowledge of the conceptual space of odors and how to judge the resemblance relations between and within an olfactory concept's instances.

The drawbacks of using high-level cognitively mediated means of identification might make it tempting to overcompensate in the opposite direction and determine odor identity in terms of stimulus detection. Mere detection might allow us to show that a gaseous stimulus is being treated as an odorant by the olfactory system and transduced as an odor, but that will only get us so far. Olfactory receptor neurons within the olfactory epithelium detect even the onset of odorless airflow (for a full coverage of the role of the motor-sensory systems in relation to smell, see Young, 2017). And whatever smells are, they must smell![8]

What is needed is a means of determining how token odors can be individuated and demarcated as a smell of a certain kind discrete from others. Thus, it is arguably the case that perceptual discrimination is the most viable measure for generating odor identities (Keller, 2017; Young et al., 2014). Our ability to discriminate between serially presented odorants provides the most accurate method for generating odor identities, since there must be some qualitative character about the stimuli in light of which the subject is able to judge them as being sensorily distinct. Using discriminability as the primary means of determining olfactory quality identification has the further strength of allowing us to determine odor identity independent of conscious awareness. Most of our olfactory perceptual states occur in the absence of conscious awareness of the odor and our attention to the odorant plumes traversing our nostrils (Yeshurun & Sobel, 2010; Young, 2014; chapter 6). Methodologically, determining odor identity in terms of an individual's judgments of just-noticeable differences (JNDs) between odorants provides the most promising method for identifying the olfactory quality of a smell because it allows a means of determining odor identity independently of conscious awareness, as well as a means for accommodating individual differences.[9]

Embracing perceptual discrimination (using olfactory quality) between odorants as the primary means of determining odor identity doesn't yet provide an explanation of what generates an odorant's olfactory quality because it only supplies us with the means by which we can judge that a given odorant has an odor that is different qualitatively for the subject from another odorant. Nevertheless, this starting point sets up the further question of what it is about the discriminable properties of an odorant from which olfactory quality arises. Considering odorants in terms of their properties as chemical stimuli provides the next step in uncovering the determinates of olfactory quality based on their material constitution.

We can specify the set of possible odors, for human beings, by noting the size of odorant molecules—not larger than twenty chemical groups and no smaller than three—that are biologically detectable by our olfactory system.[10] The general requirements for a material object to be classed as an odorant that we can perceive as having an odor is that it should be volatile and hydrophobic and have a molecular weight of<~300 daltons (Ohloff, 1986). Specifying the determinates of olfactory quality based on an odorant's chemical character suggests the molecular structure theory, which claims that what we perceive as odors are the molecular structure of chemical compounds within odor plumes. What we smell is the three-dimensional structure of molecular compounds as formed by their constituent chemical groups. Empirical evidence supporting this claim is provided by the leading scientific theories of primary olfactory transduction, our physiological olfactory capacities, psychological olfactory abilities, and animal models of olfaction (Young, 2016, 2019a, 2020).[11] The arguments and evidence that follow are supported by chemosensory findings. Yet, the precise structural properties responsible for determining olfactory quality are left open to be settled by the empirical sciences. The central tenet of MST is that olfactory quality can be primarily accounted for in terms of the molecular structure of chemical compounds within the odor plumes. Evidence for this assertion based on a simple chemical structure of odorants are enantiomers—that is, molecular compounds whose structure and functional groups are identical but whose chiral properties (i.e., handedness) differ, such that the molecules cannot be symmetrically superimposed on top of each other. By a slight majority of the tested enantiomers, individuals cannot detect different versions of an enantiomer and mostly perceived them as having the same olfactory quality.[12] But a large range of enantiomers are perceived as having different olfactory qualities. For example, R-carvone smells minty, while S-carvone smells like caraway (Boelens & van Gemert, 1993). Thus, when a subject can perceptually detect the difference between enantiomers, the symmetry of the functional groups and the orientation of a molecular compound appear responsible for a difference in perceived olfactory quality.

The existence of enantiomers that smell different provides evidence for MST. However, such cases are by a small margin not the majority. One explanation of why smelling such pairs does not always yield two olfactory qualities is based on sensory insensitivity, such that large portions of the population cannot distinguish between the majority of enantiomers because they cannot detect at least one of any given pair. Besides attributing the lack of qualitative difference of enantiomer pairs to lacking the necessary receptivity, there is also evidence that we can be trained to distinguish between previously indistinguishable enantiomers. Research on olfactory sensory sensitivity, using classical conditioning, has demonstrated that enantiomers can be distinguished by subjects, despite their original reports that they smell the same (Li et al., 2008). Li et al.'s results suggest that while optical isomers are supraliminally indistinguishable, the two enantiomers can be distinguished after classical conditioning, which can be interpreted as evidence that there are qualitative differences between the two types of olfactory experience.

In addition to our ability to discriminate between enantiomers, we can distinguish between aldehydes—that is, compounds that differ in one carbon group (Imamura et al., 1992). This nicely demonstrates our sensitivity to changes in functional groups and is intriguing because the olfactory quality of a compound changes as carbon groups are added (Turin, 1996, 2006). The aldehydes from C8 to C12 all display an interesting shift in smell as each carbon group is added; those with an even number of carbon groups smell fruity, while those with an odd number have a floral waxy odor (Arctander, 1994).

Further evidence that the molecular structure of chemical compounds yields olfactory quality is provided by our perception of functional groups (Klopping, 1971). Functional groups are atomic groups within a molecule that account for its chemical properties and structure. Interchanging functional groups within chemical compounds often creates a predictable change in the qualitative character of odorants (Turin & Yoshii, 2002, p. 11). MST fits with these results, since the structure of a molecule's chemical composition determines its characteristic olfactory quality in a predictable fashion.

Molecular structure might be the primary determinate of olfactory quality, but the concentration of the odorants within the odor plume also determines olfactory quality in a fashion that requires taking the constituent structure of the plume's token odorants and their ratio and concentration into account (Young et al., 2020). In some binary mixtures, changes in concentration yield perceived differences in odor quality (Asahina et al., 2009; Malnic et al., 1999; McNamara et al., 2007; Pause et al., 1997). However, shifts in olfactory quality as brought about by changes in concentration levels are the exception and not the norm (Cleland et al., 2012; Gross-Isseroff & Lancet, 1988; Uchida & Mainen, 2007). Learned concentration invariance[13] and the general lack of shifts in olfactory quality as brought about by changes in concentration levels explain the dearth of literature. Nevertheless, humans use a larger set of descriptions of odor qualities for monomolecular structures with greater structural complexity (Kermen et al., 2011), which can be taken as evidence that molecular complexity plays a causal role in generating the reported odor quality even in the absence of additional odorants.

While the molecular structure of the chemical compounds composing the odor are claimed to determine olfactory quality, this is, of course, against the background of individual ORN receptivity and the properties of the stimulus as a distal entity within a gaseous plume. Because of receptor genetics differences, individuals will have dissimilar receptivity to the same range of odorants. Genetic variance generates ORNs with alternative sensitivity ranges. Thus, individuals perceiving the same odorant will experience it as having different qualities. By relativizing odor quality identification based on epistemic concerns to the receptivity of the olfactory system, odor quality variance is expected and predicted.

Despite MST drawing upon a long history of research on stimulus odor relations (SOR) (Rossiter, 1996), primary sensory transduction within olfaction is still a black box.[14] So, I will leave the exact nature of the structural properties of the stimulus responsible for determining olfactory quality as an open matter. SOR is certainly a useful framework for supporting MST. However, it is important to note that MST does not claim that olfactory quality is not wholly determined by the olfactory stimulus in a causal feed-forward fashion. The olfactory quality of a smell is not simply determined just by the odorants' material properties as transduced at the receptor cite based upon some manner of an isomorphic mapping relation between chemical composition and odor qualities.[15] We should not expect the chemical properties of the odorants responsible for odor quality to correspond directly with our expectations from the methods of organic chemistry, as the kinetics of odor stimuli depend on both physical and chemical parameters. In fact, the relevant properties might be determined based on biosteres relative to the function of sensory neuron receptivity (Tahirova et al., 2019). Additionally, it might be more fruitful to look at SOR in terms of a multiple-to-multiple relational structure that is also determined based on the psychochemical properties of the stimulus in combination with the use of descriptors for the perceived odor (Licon et al., 2019).

The merit of MST is that it can be used as the starting point for further explanations in providing the determinates of olfactory mixtures and concentration invariance, as well as how we identify smells as odor objects, and perceive these against a background array of odors within a smellscape. Maybe the initial starting point of the theory is underdetermined from the evidence, and future research is required to provide validity, but its reliability might be judged by its explanatory value in generating a comprehensive account of what smells are. However, everything that has been claimed thus far depends upon only considering cases of isolated types of singular odorants. Yet, naturally occurring smells are composed from a large number of different types of odorants that we learn to group together into mereologically complex, changing, and repeatable persisting odorous objects within the environment. Thus, the next few sections will focus on fleshing out the comprehensiveness of my account by considering olfactory mixtures, how the object of olfactory perception as smells must encompass properties beyond olfactory quality, and, lastly, how we distally perceive smells as particulars within smellscapes.

2.1.2 Complex Mixtures (Question 1: Next Steps)

Initially, MST was developed with a focus on single odorants (Young, 2016). However, in generating an account of the olfactory quality of smells, even with this limited range of stimuli, it becomes apparent that the odor plume also plays a role in determining olfactory quality. Thus, the theory was expanded to handle the distal odor plume and our perception thereof (Young, 2019a, 2019b, 2020; Young et al., 2020). However, further research is required in studying how the odorant composition of an odor plume modulates our perception of the resultant smell's olfactory quality.

In particular, we need an account of the determinates of olfactory quality for mixtures, given that naturally occurring odorant plumes are composed from more than one type of odorant. Our ordinary experience of odors is of mereologically complex entities formed from a host of different odorants that we somehow recognize as particular re-identifiable smells, despite shifts in their components, concentration of the whole, or even concentration ratios between the smell's constituent parts. Thus, to explain what smells are in our everyday perceptual experiences, we must account for the olfactory quality of olfactory mixtures.

Olfactory mixtures occur when two or more types of odorants are combined to yield a unified percept. In general, olfactory mixtures generate either an elemental percept, such that the components maintain their distinct olfactory qualities and are identifiable within the mixtures, or configural (synthetic) percepts, where the mixture's smell is a new olfactory quality that is not an additive combination of the component odorants. Configural mixtures are particularly fascinating, since the mixture's odor quality is not determined as an additive process, such that one cannot predict the new smell from its individual components (Berglund et al., 1973).

Initial research on rodents indicated that perceptually similar odors yield configural mixtures, while dissimilar odors yield elemental mixtures (Wiltrout et al., 2003). However, further research suggested that the resultant quality of an olfactory mixture is better accounted for by receptor sensitivity to molecular features of the odorant (Kay et al., 2005). Mixtures formed by odorants with similar molecular structures activate similar sets of receptor neurons, thereby generating configural mixtures, while those differing in structure yield elemental mixtures, suggesting that the synergistic properties attributed to the gaseous cloud might be accounted for in terms of receptor transduction and not the plume. However, there might be reason to think that gaseous plumes also play a role because similar and dissimilar components can yield both kinds of mixtures depending upon the concentration levels of the constituents (Kay et al., 2005). By varying the concentration of odorant components, one can influence whether the complex mixture will be perceived as configural or elemental (McNamara et al., 2007), suggesting that the overall gaseous object as demarcated by its concentration also plays a role in determining odor quality. Thus, this provides further evidence for MST that, even in the case of olfactory mixtures, the olfactory quality is determined in light of the molecular structure of the constituents' chemical compounds and their relations in composing the odor plume.

The distinction between configural and elemental mixtures is often treated as binary. Yet, recent studies suggest it is a continuum phenomenon, with some mixtures yielding only mildly configural odors. Both humans and rabbits treat the RC6 mixture (artificial strawberry smell) as configural (Sinding et al., 2013). However, when one of the components is removed, the mixture is treated as mildly configural, depending upon the identity of the constituent that has been removed, as well as the resultant ratio between the remaining components. Similar results have been shown using rabbit pups presented with the RC6 mixture. Surprisingly, even changing less than 50 percent of the components still yields a weak configural percept of olfactory quality (Romagny et al., 2015). The continuum of perceived odor identity between elemental and configural mixtures suggests that even properties within the composition of the mixture, aside from odorant identity and their molecular features, play a role in shifting odor quality, thereby suggesting that the odor quality of olfactory mixtures is determined through a confluence of properties of receptor transduction of the constituent odorants together with further properties of how the olfactory system encodes and represents complex smells.

The olfactory quality of elemental mixtures might be explained as the combination of intrinsic qualities as determined by the sensory qualities intrinsic to the structural properties of matter that the olfactory system transduces, but in explaining configural mixtures, such an account seems dubious because the smell is not the sum of primitive's constituent odors. While something as simple as discriminability between the chemical structure of odorants might explain the olfactory quality for singular types of odorants, when olfactory mixtures come into perspective, it becomes clear that more complex process will need to be adverted to then simple feedforward systems (FFS) of stimulus transduction. How we encode collections of different kinds of odorants as singular identifiable odors requires noting how the system representationally encodes olfactory mixtures, how cortical process represent complex odors, and how we perceive and think about smells as objectual categorical entities. More will be said about this in chapter 4, which shows that the olfactory system employs a combinatorial system of representation that is non-compositional in the classical sense.

However, for the purposes of supporting the claim that olfactory quality must also be accounted for representationally, then the format employed by the system is a factor as well.

Although it was commonly thought that odorants were coded in a coarse manner at the receptor and olfactory bulb (Asahina et al., 2009; Friedrich & Laurent, 2001), Vincis et al. (2012) showed that these results are attributable to the odorant and the anesthetized state of the organism. Under natural conditions using ordinary odorants, they recorded robust fine-grained representations within the glomeruli of the OB in mice. Despite these results, several studies using rodents show completion effects whereby, in multicomponent mixtures, the olfactory system shows either the same coding, despite the absence of a constituent, or a change to the constituents. For instance, Johnson et al. (2010) showed that olfactory coding within the OB represents the major molecular features of the stimulus. Yet, for some complex mixtures, the coding was less complex than would be expected if all the molecular features were represented, which indicates that only major constituents of the complex are being represented. Their results confirm previous findings (Johnson & Leon, 2007) that the encoding of molecular features primarily maps the major components of an olfactory mixture. Similar findings regarding the representational format of complex smells within the human piriform cortex suggest that we often represent complex smells in a holistic fashion, such that the olfactory quality of some olfactory mixtures requires quantifying over high-level olfactory processes and cognitive states relative to the unique format of smells (see chapter 4 for a more in-depth discussion of evidence supporting these claims).

However, once it is allowed that olfactory quality is not simply a matter of SOR and feedforward processing of transducing the odorants but might also encompass cortical encoding and cognitive states, then this leaves room for individual difference in subjective reports of the olfactory quality based on either differences in receptivity or learned associations. Naturally, if our experience of odors, as identified in terms of their olfactory quality, depends on how the olfactory systems generate our representation of complex stimuli, then there is room for wondering how it is that we experientially represent smells—that is, what the intentional object of smell is. Thus, the next section turns to considering if we need to posit olfactory objects in explaining our perceptual experiences and cognitive states about smells.

2.2 Question 2: What Are the Odorous Objects Represented within Our Olfactory Experiences?

The olfactory system allows us not only to detect the presence of odorants and recognize ecologically relevant changes within the chemical sea we are immersed within (Barwich, 2020; Keller, 2017) but also to perceive odors as perceptible objects with complex mereological structures with changeable and repeatable sensory properties across their perceptual instances (Young, 2019a, 2019b, 2020). From the outset, it should be noted that, in discussing odor objects, it is important to distinguish between what the distal odor objects are that we transduce and encode using the olfactory system and what the objects are that we experience as smells. The former issue concerns the distal nature of smells as particulars and how we perceive these against a background array of odors, which is handled in section 2.3, while the latter is the focus of this section concerning what I have previously referred to as the intentional object of smell and herein rephrased as question 2: What are the odorous objects represented within our olfactory experiences?

Reasons for positing odor objects include philosophical arguments that olfactory experiences are representational (Batty, 2010c; Lycan, 2000, 2014; Young, 2016), have a figure–ground structure of perception (Millar, 2017; Stevenson, 2014; Young, 2016), are experienced as mereologically complex entities (Millar, 2017, Skrzypulec, 2019; Young, 2020),[16] can solve the many-property problem (MPP; Young, 2019b), and, according to some objectual theorist, can even generate perceptual experience with amodal completion (Young & Nanay, 2021).[17] With all that said, there are strong arguments supporting the conclusion that olfactory perceptual states generate complex experience that organize disparate sets of odorant stimuli into unified objects of perception with changing and repeatable sensory properties.[18]

Rather than repeat and review the robust literature concerning olfactory objects (see Batty & Young, forthcoming), this section progresses the debate by offering further reason to think that the olfactory object should be considered not just in light of the olfactory quality responsible for generating the primary means of odor identification but also as a host of complex and interrelated sensory properties that we experience a smell as having. Smells are experienced as objects of perception beyond their olfactory quality. We also experience smells as being intense, concentrated, and pleasant or disgusting. My previous arguments that we must consider smells as complex mereological entitles (Young, 2016, 2019a, 2019b, 2020) that built upon O'Callaghan's (2007, 2008) theory of auditory objects and the ecological approach to olfactory objects (Gottfried, 2010; D. A. Wilson & Stevenson, 2006) must be expanded to account for these further changing and repeatable olfactory properties. While smells must smell (i.e., have an olfactory quality), they are far more complex than this trivial platitude admits, such that the sensory space for our experience of odors includes separate dimensions for odor quality, intensity, valence, and thresholds for odorant detection (Bierling et al., 2021).

According to MST, the primary determinate of odor identity is its olfactory quality that is accounted for by the structural properties of the odorant stimulus in composing odor plumes. However, in accounting for how we perceive smells as objects, the theory must be expanded to account for our experience of the smell's concentration, intensity, valence, and other associated properties. We represent smells as re-identifiable particulars with a holistic complex nature that allows us to perceive them as persisting objects across their changes in distinct sensory properties and instances of perception. Explaining the holistic yet complex representation format of smells will be left aside until chapter 4, but we can explain the disparities in experimental findings about the primary determinate of smells if we note that smells are complex perceptual objects including olfactory quality, concentration, the composition of a plume (as noted earlier), as well as their valence and intensity.

2.2.1 Smell's Valence

Despite the evidence that the molecular structures of odorants allow us to individuate an odor as having the same sensory quality across perceptual instances, it is still debatable whether the primary determinate of the identity of an odor (odor identity) is its valence (being a pleasant or unpleasant scent) or its olfactory quality.[19] In a series of studies, it has been argued that valence is the perceptible property used by humans to determine odor identity (Yeshurun & Sobel, 2010). Unlike odor-quality categorization, which is similar in various respects but varies cross-culturally, there is greater agreement on the categorization and identification of odors using judged pleasantness or unpleasantness (Haddad et al., 2008, 2010). Moreover, Snitz et al. (2013) generated a computational model of odorants that can predict perceived valence from their chemical structure alone.

However, competing research indicates that humans more likely identify the perceived identity of an odor in terms of its odorous quality (Olofsson et al., 2012, Olofssson, Bowman et al., 2013). Additionally, Kumar et al. (2015) created an alternative computational model to that of Snitz et al. (2013) using descriptors of odor qualities and not judgments of valence, as well as measures of chemical structures to predict olfactory quality. Similarly, Mantel et al.'s (2019) test for human olfactory change detection only yielded reliable detection for 24 percent of the participants. Yet, across all individuals, olfactory quality was detected with greater frequency than concentration, suggesting that odor quality is primary for the purposes of odor identity.

Not only do behavioral tests support the claim that our identification of smells might require taking the valence of the odorant into account in addition to its odor quality, but further research on cortical processing has also shown synchronization of olfactory processing centers differs between tracking the hedonics and the quality of a smell. Additionally, there seem to be separate cortical processing centers within the piriform cortex for representing odor identity, valence, and intensity (for a review, see Blazing & Franks, 2020). What is especially noteworthy in favor of MST is that the molecular features of an odorant are correlated with its perceptual qualities in a manner that allows for separable dimensions of quality, pleasantness, and intensity (Keller & Vosshall, 2016).

2.2.2 Smell's Intensity

Intensity is often oversimplified as the strength of the quality of smell and is often conflated with concentration, which is the density of molecules of a particular odorant within a spatiotemporal boundary. While the determinate of odor intensity is still being researched, it seems plausible that a smell's intensity is dissociable from its concentration. Nevertheless, the intensity of an odor is linked to the concentration of odorants composing the odor plume.[20] Yet, odor quality, valence, and intensity are perceptually dissociable and might be determined by different sensory mechanisms. Research on mice further supports the claim that odor quality and intensity are dissociable, since mice display distinct cortical coding strategies for odor identity and intensity (Bolding & Franks, 2017). For a good overview of the dissociation of odor quality from intensity and a theory of how odor intensity might be determined from the concentration gradient of a complex odor plume, see Mainland et al. (2014). Similarly, Giaffar et al. (2018)

generated a primacy model of odor identity with respect to intensity that shows how initial ORNs encode odor identity with respect to quality across a range of different concentrations. Giaffar et al.'s model generated testable predictions of how a small set of high-affinity receptors could encode odor identity within a single sniff, with the focus on odor identity as determined by intensity and not quality. Thus, if chemoscientific research eventually validates these predictions and if models such as those developed by Giaffar et al. and by Snitz et al. (2013) continue to challenge the primacy of odor quality, then my own approach might need to be modulated to allow for multiple means of determining a smell's identity.

What is of further interest is that these sensory properties of smell are highly interrelated, such that there is a negative correlation between intensity and valence (strong smells are often judged as unpleasant), and judged salience and arousal of a smell are correlated with its valence. It is also often the case that odor quality and valence overlap in terms of the structural properties of the stimulus. For a fuller discussion of the dissociation of these dimensions and their interrelation within the odor space of our experience, see Bierling et al. (2021). Suffice it to say that what we perceptually experience as a smell extends beyond its odor quality.

Arguably, smell's primary identity is as a qualitative smelling thing, which sets it apart to be objectified and individuated from other smells. Once the smell is identified in terms of its quality, we also ascribe to it further qualities of valence, concentration (density), and being strong or weak (intensity). All of these further properties can shift across token instances of a smell once it is identified as falling under a particular type, given its olfactory quality. Our experience of smells admits to being as of a multifaceted, mereologically complex perceptual object, but how this is experienced as a distal object and the spatiotemporal properties thereof is a matter to which we now turn.

2.3 Question 3: What Are the Distal Entities That We Perceive as Smells?

Our perceptual experience of smells and olfactory capacities extend beyond tracking odorant mixtures, their olfactory qualities, and corresponding properties. We also smell odors as extended particulars within smellscapes. Despite the introspective phenomenology of our smell experience reporting that smells merely exist within our environment or appear as transiting our nostrils,[21] olfaction is well adapted to track complex olfactory mixtures that are of salience for us for ecological purposes within and across an environment.[22]

This section will focus on briefly summarizing the difference between nonobjectivists' accounts of smells and my own.[23] In particular, it will be argued that both Batty's and Keller's view shares a similar set of premises about the spatial nature of smells that I find dubious. Arguably, it is the case that their accounts depend upon certain methodological differences. I will highlight these and explain what I find questionable about these starting points. The section will end with a synopsis of why I think we should consider smells as mereologically complex distal entities that can be perceived within and against the background of a smellscape by noting that what we consider the objects of a given modality should be sensitive to the spatial and temporal relation of the perceiver to the object of perception of that modality.

Batty develops the abstract account of smells over the course of multiple arguments concerning the veridicality of odor perception, the individuation of multiple odors within an array, and the many-property problem. As an odor theorist, the object of olfactory experience is a gaseous cloud of odorants. Yet, the object of olfactory experience according to Batty is not of a particular entity composed by a gaseous cloud. Rather, we experience an odor as a property of the environment. According to Batty, correctly describing the representational nature of our olfactory experiences requires making use of an existential quantification that there exists some smell hereabouts in our environment, which is not to claim that odors don't exist in space and don't have spatial properties but rather that they are not experienced as having a determinate spatiotemporal locus within egocentric space.

The abstract account is primarily constructed to handle the object of olfactory experience using our phenomenological experience of the distal nature of smells as presented to us (almost) instantaneously within conscious self-introspective reports. However, Batty's focus on the phenomenology of these temporally punctuated experiences of smells modeled on the timescale of visual perception generates the overarching claim that olfaction does not present locatable entities with fixed spatial locations. While the theory concedes that olfactory experiences have spatial aspects to them, the olfactory object is not identified in terms of spatial properties.

The theory provides a strong explanation of the object of olfactory experience, but it does not provide an explanation of the olfactory quality of smells (Young, 2016) or an account of our distal perception of smells across time (Young, 2019a, 2019b).

Keller's *Philosophy of Olfactory Perception* (2017) provides a rather exhaustive treatment of the philosophy of olfaction, centering around the claim that olfaction is designed for the determination of behavioral output. Accordingly, olfactory perception should not be conceived of as generating accurate representations of external chemical stimuli because the functional role of olfaction as a sense is the detection of salient entities for behavioral output. Regarding the distal nature of smells, Keller argues that olfactory qualities do not have spatial properties because we do not have spatial properties presented to us as part of the olfactory quality. In those instances where it seems that olfaction provides us with spatial information, he rightly notes that these might be attributed to chemothesis. However, his arguments might be questioned based on his narrow construal of smells merely in terms of olfactory quality. If smells are conceived of as complex mereological entities identified in terms of their olfactory quality with a host of other properties, then his conclusions seem less convincing.[24]

Despite offering independent theories of olfactory experience and perception, Batty's (2010a, 2010b, 2010c, 2014, 2015) and Keller's (2017) arguments that smells should not be considered objective entities within the environment are predicate upon a similar two-part argument, whose conclusion is that olfaction cannot resolve the many-property problem.[25] Batty and Keller's shared argument might be summarized as follows: olfactory perception and/or experience does not present olfactory objects with fixed spatial locations, such that it cannot generate figure–ground segregation of an odor array. Thus, it cannot resolve the MPP. However, the argument depends upon a short (synchronic) temporal time frame of perception garnered from how the phenomenology seems within introspective self-reports of perceptual episodes of distal olfactory perception, whose temporal constraints are derived by analogy from vision. With this methodological access point to olfactory perception and experience, their theories generate a truncated olfactory perspective that limits what might be considered an olfactory object (for a thorough review of these arguments and criticisms, see Young, 2019b).

Batty and Keller offer four arguments for the conclusion that we do not (synchronically) perceive and/or experience smells as being spatial entities within the environment based on our poor spatial resolution, our inability to discriminate between and individuate olfactory particulars within an overlapping array, our inability to detect odor onset relative to nostril onset, and the lack of intrinsic spatial properties within the experience of olfactory qualities (reviewed in Young, 2019a). Each of these arguments shares the underlying claims that we do not experience and perceive smells as individual entities with set boundaries and fixed spatial coordinates in the environment.

If olfaction does not present perceptible objects at specifiable coordinates in allocentric or egocentric space within their temporal perspectival constraints, then it becomes uncertain if olfaction could generate the capacity to segregate one perceptible object from an array of background entities. However, it is not clear that we should grant the initial starting point and methodology that extrapolates the temporally short olfactory perspective by analogy from vision. As noted previously (Young, 2019a), the methodology of inferring from our phenomenological reports of visual experience to that of olfaction might be questionable, given the underrepresentation of the background effects and influences that we report in vision. Most philosophers simply assume that, upon opening our eyes, we are immediately presented with an array of objects with punctuated boundaries against a background. Yet, there are reasons to doubt the methodology of using introspective access in general because it is an unreliable mechanism that provides unreliable results (Schwitzgebel, 2008). Moreover, it is arguably the case that there is a multiplicity of introspective mechanisms that draw upon our background knowledge and cross-modal integration (Schwitzgebel, 2012). In particular, these errors can be seen in our accounts of visual experience that pay no heed to visual object perception requiring years of development, cross-modal integration, and a profound amount of information shared from cognitive states. The dubious utility of using phenomenological reports of experience gained through introspective access is even more pronounced when applied to olfaction. The phenomenological method of assessing the nature of olfactory experience might be both unreliable and invalid as an accurate means of assessing the intentional and distal object of olfactory experience. The vast majority of olfactory experiences occur in the absence of attention, thus bringing into question if phenomenological reports of consciously aware experiences are a reliable guide to the nature of the representational format of these experiences, given the possible bias of using a suboptimal sample set (Sela & Sobel, 2010). I think these concerns provide strong reasons to doubt the veracity of Batty's and Keller's claim regarding the spatiotemporal nature of distal olfactory perception, but I don't want to rule out that perhaps some smell perceptual experiences do phenomenologically present themselves in the truncated (synchronic) perspective they describe. These are, however, outliers.

When we consider smell perception under both experimental and naturalistic conditions, which correctly account for the olfactory system's extended temporal sequences of odorant transduction and processing, a different conclusion for question 3 might be warranted. Accounting for olfactory perception requires being sensitive to both the spatial aspects of distal perception and the temporal characteristics of olfactory processing. If olfactory temporal processing is slow and cannot be determined according to the common timescale of visual object perception, then, arguably, our ability for spatial olfactory perception should follow suit. Smelling odors within an environment is an extended process. We cannot demarcate odors as occurring at a given place with the same time frame as vision. Rather, we locate smells as occurring within an environment against the background of other odors across time.

Olfactory perception is rather slow. The average sniff lasts 1.6 seconds. During the initial phase of sniffing, we modulate the volume of airflow, pressure of airflow, and sampling rates. Additionally, toward the middle to end of a sniff, we can detect the presence of an odor, as well as identify its olfactory quality (what it smells likes) and valence (reviewed in Olofsson, 2014). The sniff sequence can be segmented into multiple stages. The initial sniff onset brings the stimulus into the nasal cavity and lasts 200 milliseconds. Within 150–300 milliseconds of stimulus presentation, sniffing is modulated in accordance with the concentration, intensity, and valence of the odorant. Additionally, within 150 milliseconds of sniff onset, we modify or sniff response in accordance with the olfactory valence of the stimulus. Encoding the olfactory properties of the odor occurs during a 500-millisecond period following the initial 200 milliseconds of sniff onset. Only after 800 milliseconds of sniff onset do we consciously detect the odorant. Identification of olfactory quality and odor valence follows at intervals of approximately 1,000 and 1,100–1,200 milliseconds, respectively (reviewed in Olofsson, 2014; summarized in Young, 2014, 2020). Thus, careful attention to the temporally extended nature of olfactory perception calls into question holding the spatial nature of olfactory objects fixed to the perspectival relation we extrapolate from vision. Our olfactory experiences, when held to the temporal constraints of vision, do not present odors as particulars with spatial properties within an array, but once we jettison such comparisons, we can appreciate that olfaction's slow temporal processing speeds allow us to parse olfactory mixtures diachronically as mereologically complex repeatable entities within an overlapping scene of alternating smells (Young, 2020).

Initially, MST was developed to handle single odorants (Young, 2016). However, the theory was expanded to handle the distal odor plume and our perception thereof (Young, 2020; Young et al., 2020). To gain a better grasp of how the odorant composition of an odor plume affects how we perceive olfactory quality, the experimental literature was reviewed across animal models concerning odor plume perception, olfactory navigation, and odor tracking (Young et al., 2020). Based on this review, it was concluded that, in accounting for the molecular basis of smells and our distal perception of odor plumes, we must quantify the odor plume as a superordinate perceptual object beyond its mere constituent odorants and the concentration gradient of its filaments. Thus, based on the review, it can be concluded that not only do we perceive smells as distal entities, but we must also advert to the plume of an odor as a distal object within a perceptual array of overlapping and occluding smells.[26]

We perceive a smell as an odor plume encompassed within an olfactory scene conceived of as a smellscape. In accounting for the distal object of olfactory perception as particular smells with an extended odorant plume, we need to allow that our introspective phenomenology often does not present smells as distal entitles. However, our retrospective reports of olfactory phenomenology might be inaccurate when accounting for more naturalistic conditions in which we perceive smells as expansive entities that we segmented as particulars from the surrounding sea of gaseous odorant plumes within which we are constantly immersed.

Smelling objects within an environment takes time. We cannot demarcate odors as occurring at a given location within the same short timescale in which we can localize visual objects, but we are able to locate smells as occurring within an environment against the background of other odors across time. Noting that the temporal processing of the olfactory stimulus is extended provides motivation for considering our ability to locate and perceive odors as spatially extended. Allowing a spatial expansion of the perceptual scene suggests that the distal object of olfactory perception will have to be of a spatiotemporally extended array (Young, 2020). Our intuitive theoretical launchpad employing vision to theorize about distal object perception is misguided when applied to smells. Rather, we need to rethink the nature of perceiving at a distance relative to the object of perception for each modality. We perceive smells as perduring mereologically complex odors that are distal objects with spatial properties within an olfactory array that might be conceived by analogy to landscapes as smellscapes. However, claiming that the identity criteria of distal objecthood must be determined based on the transduction process of sensory systems relative to perspectival relation between the perceiver and distal object of perception must be relativized to a modality. As such, MST takes a rather strong stance on the sensory systems that constituted our sense of smell. While smells might have perceptible aspects beyond olfactory quality such as their valence, intensity, and spatiotemporal properties, according to MST, the primary means of determining odor identity is the olfactory quality. Thus, not every sensory system within the nose can be considered as part of our sense of smell—if it doesn't detect olfactory quality, then it isn't part of smell.

2.4 Conclusion

There is a seeming incompatibility between my account of olfactory quality identification and olfactory perception of smells as distal entities. Smells are primarily identified by olfactory quality, but in case of distal olfactory perception, it would seem that a smell's identity includes properties of the odorants and process beyond those that are essential for transducing olfactory quality. Elsewhere, I have argued that using olfactory quality as the means of individuating the perceptual modality of smell rules out trigeminal and somatosensory systems within the nose from the sense of smell (Young, 2017), and olfactory perception is relative only to olfactory quality and the mechanism responsible for olfactory quality (Young, 2020). Yet, when expanding the perceptual time frame for naturalistic conditions to encompass the perception of complex perduring odors within a smellscape, we must include these very same systems in accounting for the exploratory movements required to perceive smellscapes. Put succinctly, it seems that I exclude these processes from the perceptual modality of smell for olfactory quality perception yet consider them components for our perception of smells as distal entities.

Other sensory systems play a role in our perception of the spatial extent of the smells within a smellscape and our perception of the odor plume's filament structure. But these are best conceived of as background-enabling conditions required to perceive the spatiotemporal properties of the odorant plume but which are not constitutive of the olfactory system. Similarly, many background processes, including saccadic eye movements, are essential for our visual capacity to see objects. Yet, we ordinarily exclude such necessary eye and head movements from being constitutive of the visual modality. Sniffing, exploratory bodily movement, somatosensory sensations inside our nose and on our face, trigeminal stimulations, and respiratory patterns all play an influential role in our capacity to perceive smells as distal objects in terms of their tracking properties of the odorant plume. But it would be a mistake to accord these processes status as part of the olfactory modality, since they only associatively generate experiences of olfactory quality and often track structural properties of chemical stimuli that are odorless (Young, 2017, 2020). Moreover, in their role as supporting sensory systems, they often generate different types of sensory qualities in combination with other perceptual modalities.[27] Each of the aforementioned nonolfactory sensory systems are best treated as having constitutive roles in generating our perception of smells as complex entities with a variety of perceptual features beyond their olfactory quality, including associative, conceptual, and other cognitive states.

There is an outstanding issue here regarding the conceptualization of background-enabling sensory systems that play constitutive roles for a perceptual system yet are not considered part of the modality. How to individuate the senses is a perennial philosophical problem that is not specifically more pressing for olfaction. And of course, this onus of proof-style theoretical ping-pong is not very satisfying as a reply. Yet, I have provided arguments for requiring these other systems as constitutive elements that enable our capacities to demarcate and individuate smells as distal objects, as well as reasons to exclude these from the modality of smell, because odor identity derives from the olfactory quality of the stimulus (Young, 2016, 2017, 2019a, 2019b, 2020). The further issue of how to individuate senses given background constitutive sensory systems certainly requires further research, especially in relation to the chemosenses, as noted previously in Young (2020) and Young and Nanay (2021).

One way that this issue could be tackled that fits MST's claims that smells essentially must smell, that the olfactory object is experienced as a mereologically complex entity with a range of sensory properties, and that orthonasal smell and flavor are distinct would be by adapting Skrzypulec's (2021, 2023) framework for demarcating constitutive from merely influential sensory component of flavor experience. His fine-grained distinctions between types of constituencies and, in particular, minimal ones that are sensitive to underlying sensory mechanisms could be employed in cleaving smell individuated in terms of its olfactory quality from the other sensory systems that enable our encoding of the properties of odorant plumes enabling our distal perception of smellscapes. However, at this time, I do not have an explicit position that is superior to my speculative claims above regarding platitudes about the primacy of olfactory quality in determining odor identity and the essential nature of smell. Forgive my holding some projects back for future research. For now, I hope the coherence of my answers to questions 1–3 provides a satisfactory explanation of what smells are.

Answering the three nested questions of what accounts for a smell's olfactory quality, what is the object of olfactory experiences, and what is the distal object of smell perception provides a rather interesting ramification that what we consider to be smells is only relative to orthonasal olfaction. Our experience of smells derives from inhaling olfactory stimulants from the front of the nostrils, while retronasal olfaction, when odorants are pushed upward from the mouth through the throat and exhaled out of the nose, is not part of what we consider the sense of smell. While this seems like a rather drastic claim, it is not only in keeping with our conception of smell, but also arguably the case based on the differences between how the two olfactory pathways differ in their olfactory qualities of the same stimuli (Barwich, 2020; Young, 2023), how they transduce complex smells, the intentional object of each pathway, and the nature of their distal perceptible objects (Young, 2023). However, if we don't retronasally perceive smells, then what are we perceiving with this separate olfactory pathway?

3 Tasting Flavors

The ebb and flow of our average day through its transitions between sleep and wakefulness, work and home life, and responsibilities and relaxation revolves around the consumption of food and drink. Whether this daily schedule of meals and snacks is implicit or explicit, it is hard to shake the feeling that our life is oriented toward eating and drinking. We are drawn to the consumption of flavorful entities. For instance, the gravitational pull of cravings demands specific types of items and their exact sensory qualities. Mere similarities in the flavor profile of a consumed interloper will not satisfy the craving. We replicate family recipes, purchase particular beverages, and plan trips to specific establishments to order the same dish in the hopes of reacquainting ourselves with beloved sensory experiences of flavorful items. Despite being dominantly visual beings, it is hard not to appreciate the force that our drive for tasting flavors exerts over our life. Yet, the nature of taste and flavor perception are not as well understood. Thus, the purpose of this chapter is to argue for an account of our perception of flavor that also allows for an expansion of the perceptual modality we intuitively consider as taste as including a host of multisensory transduction channels that together form a uniform perception of flavorful entities.

The question of how to individuate the senses has recently enjoyed a resurgence in philosophy (Macpherson, 2011a). The current interest in the senses coupled with the popularity of research on multisensory and multimodal perception have thrust olfaction and flavor perception into the spotlight (Fulkerson, 2014; Macpherson, 2011b; Matthen, 2015, 2017; O'Callaghan, 2015, 2016; B. C. Smith, 2009, 2013, 2015). However, the few discussions of the relation between retronasal olfaction, taste, and flavor (Macpherson, 2011b; Matthen, 2015, 2017; O'Callaghan, 2015; Richardson,

2013a; B. C. Smith, 2015; K.A. Wilson, 2021) have not generated any consensus on the relationship between taste and flavor (Spence et al., 2015). Yet, by adopting Fulkerson's (2014) pluralist approach for individuating the senses, such that what we consider a perceptual modality depends upon our explanatory purposes, it is argued that, at least for the chemosenses, one explanatory strategy is preferable that cleaves these senses apart based on their object of perception (intentional and distal), their sensory qualities, and the format of perception As argued in the last chapter, at least for smells, a comprehensive account of the nature of smell requires handling the nested issues of accounting for (1) olfactory sensory qualities, (2) the object of olfactory experience, and (3) the distal nature of smells (Young, 2016, 2019b). The chapter extends the threefold cluster in assessing the nature of taste and flavor perception with the addition of the representational format that is foreshadowed in the concluding section and elaborated in chapter 4.

What will be argued is that taste is not distinct from flavor perception—we taste flavors. Flavors are the synthetic combination of multisensory transduction systems into a unimodal and not fully decompositional percept. Using the sensory systems within our headspace, we perceive multiphasic dynamic chemical objects—consumable entities. To support this complex conclusion, the chapter unfolds over the following sections. The first section introduces the role of retronasal olfaction in tasting and argues that retronasal olfaction plays a dominant role in taste experiences. Section 3.2 argues that taste perception should not be equated with mere gustatory stimulation. Rather, tasting is conceived of as occurring within the oral cavity, thus expanding what might be considered the organ of taste perception. Building upon this expansion, section 3.3 argues that tasting requires multiple sensory channels within our headspace, including retronasal olfaction as a constitutive element in most of our usual taste experiences. While section 3.1 is of relevance to the first of the nested issues (sensory qualities), sections 3.2 and 3.3 more directly deal with the issue of sensory qualities in terms of both retronasal olfactory qualities and taste not being equivalent to stand-alone gustatory qualities. With this background in place, sections 3.4 and 3.5 handle the second and third nested issues (the object of experience and the distal nature) by developing the claim that the object of taste are consumable entities with flavor qualities. The chapter concludes that we taste flavors, such that taste is a multisensory yet unimodal perceptual system that generates synthetic non-decomposable flavor percepts.

3.1 The Role of Retronasal Olfaction in Taste Experience

There are two distinct pathways for the odor plume to reach the olfactory epithelium. One pathway for odors to reach the olfactory system is via the front of the nose. Orthonasal olfaction, as it is so labeled, is what we primarily refer to as our sense of smell (Young, 2023).[1] Our olfactory experiences are usually of external odor plumes within the environment that have been brought into our nose through normal inhalation or actively sampling the surrounding air by sniffing. In addition to the orthonasal pathway, there is a second pathway for airflow from our mouth via the throat (nasopharynx) upward to the back of the olfactory epithelium. Most of our attended olfactory sensations are not strictly smell but rather flavor experiences. A large number of the perceptual qualities that we assign to eating are transduced by the retronasal olfactory system. The activities of chewing and swallowing cause odorant-laced air to flow from our mouths into our nose, thereby traversing the olfactory epithelium, and out of our nostrils.

It comes as a mild surprise to most people that what we ordinarily report as taste experiences are partially determined by the olfactory system. Two do-it-at-home experiments are usually employed to highlight the role of retronasal olfaction in the experience of flavor. The jelly-bean experiment[2] might be the easiest way to demonstrate the role of exhalation and retronasal olfaction in generating the experience of flavor as opposed to just the gustatory experience of basic taste qualities. Place a jelly bean in your mouth while holding your nose tightly and chew. The gustatory qualities of sweet, bitter, or sour (depending on the color of the jelly bean) should be readily perceptible. After a few chews, release the grip on your nose and exhale. What occurs is a profound experience of gaining access to the robust flavor profile of the jelly bean. It is not that we now access the additional smell qualities, but rather that we now fully taste the jelly bean. Since naive participants do not claim to gain access to olfactory qualities upon exhaling, our natural experience reveals a bifurcation of which olfactory qualities are considered smells. In analyzing the experience of the jelly-bean experiment, we do not claim that the flavor properties only exist upon exhaling. Rather, we are delighted to learn that there are some objective flavor properties of the jelly bean that can only be accessed through retronasal olfaction.

A complementary experience occurs to most of us on an annual basis. When suffering from the average winter cold including nasal congestion, the odorant plume of a flavorful soup cannot reach our olfactory epithelium, thereby causing us to experience only the gustatory taste qualities of the soup. Our congested nasal passages impede retronasal olfaction, and in these instances, we only access the most minimal qualities of the food's flavor—its gustatory qualities.

The jelly-bean and sick-soup experiences highlight how retronasal olfaction plays a role in the experience of flavor perception. However, until recently, it was not possible to deliver an odor to the retronasal olfactory system independent of other sensory channels. As Shepherd (2012) puts it, "retronasal smell is never sensed by itself" (p. 17). While this might be true under naturalistic conditions whereby a stimulus must be placed within the oral cavity, thereby stimulating somatosensory, thermal, and/or gustatory receptors, Heilmann and Hummel (2004) developed a technique for the direct stimulation of orthonasal and retronasal olfactory systems. Tubing is placed through the nasal cavity, ending either at the front of the nostrils or directly below the olfactory epithelium close to the nasopharynx. By placing the tubing in the nostrils in these two locations, a stream of odorant-laced air can be delivered directly to each olfactory system. Even in these extreme laboratory conditions, different olfactory qualities are reported based on the route of delivery. The same set of pure odorants delivered directly to each system yields differences in perceived olfactory qualities. Thus, the same odor plume will smell different to each system, making it reasonable using the first nested issue (sensory qualities) to differentiate the orthonasal perception of smells with their set of olfactory qualities from the retronasal system transduction of olfactory flavorful qualities that originate from an object placed within the mouth.[3]

Orthonasal and retronasal olfaction show further dissociations in clinical cases involving the loss of smell. Both Hummel (2008) and B. C. Smith (2015) have interesting discussions of clinical dissociations that provide reasons to think olfaction has two independent sensory systems. Furthermore, Landis et al. (2005) document eighteen patients with no complaints of taste dysfunction but who yet display a deficit in orthonasal olfaction. Using an identification task, the average subject failed the orthonasal task at 36 percent, as well as having no event-related potentials (ERPs) recorded from olfactory receptor neurons for the orthonasal presentation. In contrast, the retronasal identification test yielded 56 percent correct identification with recorded ERPs of ORNs. Based on these findings, Landis et al. argue that the two systems might not just be functionally distinct but also possibly structurally distinguishable. They hypothesize that the olfactory epithelium is structurally organized to be sensitive to orthonasal presentations versus retronasal presentations of the same set of odorants, which would both explain their clinical data and fit with research surveyed by Goldberg et al. (2018) that the olfactory quality of an odorant might depend upon the route of delivery. At least when considering the perceived sensory qualities of odorants as one explanatory criterion (i.e., sensory qualities) for differentiating the senses, we might consider orthonasal and retronasal olfaction to be independent senses. Since retronasal olfaction never occurs on its own outside of extreme laboratory conditions, it might be wondered whether, in fact, retronasal olfaction might in most instances form a sensory channel within a larger perceptual system.

It might be tempting to posit the existence of a further sense of flavor independent of taste and smell, but this would be a mistake—taste already covers retronasal sensory input. While we often refer to taste in terms of sensations on the tongue, P. Rozin (1982) showed that people do not, on average, distinguish between descriptors for flavor-based sensory qualities and gustatory qualities. Moreover, after introducing participants to the distinct role gustation plays in taste qualities and retronasal olfaction plays in flavor sensations (in English), he discovered that speakers of seven out of nine languages reported there was no semantic distinction or terminological differentiation between taste and flavor in their language. Taken at face value, despite paying lip service to the naive conception of taste occurring on the tongue, what is considered our sense of taste across languages (on average) is indeterminate between taste as gustation and flavor including retronasal olfaction.

An astute philosopher might point out at this stage that, as pluralists, we could maintain the distinction between taste as gustation and flavor including retronasal olfaction in the manner Matthen (2015, 2017) suggests, such that scientists might focus on sensory transduction systems and generate a host of different senses that do not overlap with our ordinary conceptions, whilst philosophers with their focus on unified phenomenological percepts might classify multisensory systems, such as flavor, as a unimodal perceptual system. Although I think this is partially right, there might be better and worse explanatory programs within pluralism. Just allowing for a multitude of pluralist conceptions misses that what we already consider as taste is rather broad. What we consider as taste across most cultures and languages already accommodates retronasal transduction within our headspace.

3.2 Gustation Alone Is Not Taste

What we consider to be taste experiences are thought to occur upon our tongue. Attempting to demarcate the modality of taste in terms of its organ of perception intuitively makes reference to the tongue. However, considering taste as only gustation vastly underestimates our taste experiences. The problem to be addressed within this section is whether taste should be equated with gustation.

One way to individuate the nature of taste perception is in terms of the basic qualities inherent within our perception of the taste of food placed upon our tongue. Harking back to Aristotle's individuating of the senses, we might claim that taste is the sensory system we use to perceive salt, sour, bitter, and sweet with the contemporary addition of umami.[4] The modality of taste so conceived might then be demarcated by the sensory receptors sensitive to each of these qualities, such that taste is the experience that we have when certain types of chemical stimuli interact with certain types of receptors on our tongue. Conceiving of the nature of taste experience purely in terms of the sensory organ, receptors, and basic qualities of tastants upon the tongue, then taste is equivalent to gustation.

Individuating taste perception in light of the basic taste sensations and, by extension, each of their receptor types is a viable approach that fits with what some might consider to be our naive individuation of the senses. Yet, equating the taste experience with just the gustatory experience limits the scope of what would be considered taste perception and calls into question our preconceptions about even the nature of gustatory sensations. Our intuition is that there are different sensor types, each sensitive to different types of qualities that correspond to these basic tastants. Yet, recent research suggests this isomorphic line–line system conception of the sensory encoding of basic tastes is not accurate. We have multiple sensory-chemical channels and receptors for transducing a single basic tastant. For instance, there are multiple receptors capable of transducing a range of stimuli yielding bitter and sweet experiences (for a good introductory review, see Stevenson, 2009a, pp. 13–14). Moreover, a large portion of the population believes the myth that the different basic tastes with their respective receptor types are distributed unequally into quadrants along the front, back, middle, and sides of the surface of the tongue. There does seem to be some differential sensitivity to sweetness in terms of overall thresholds between the front and back of the tongue. However, these differences for sweet sensitivity might be attributed to there being a greater number and density of sensory receptors on the front of the tongue.

We claim that we taste with our tongue. However, interpreted literally, this view of the sensory modalities will cull most of the qualities that we describe as part of our experience of tasting food or drink. Most taste experiences are not simply generated by the gustatory system. Moreover, taste receptors are located throughout the lining of the mouth. Arguably, a more charitable interpretation would be that we taste things in our mouths. Additionally, the aforementioned study by P. Rozin (1982) suggests that even speakers of outlier languages such as English do not differentiate between gustatory and flavor descriptors when talking about tastes. Thus, taste seems to be more expansive than just gustation when considering the sensory organ of taste, the sensory qualities of taste/flavor, and where we locate the perceptible object (that might go some way to demarcating the distal nature of tasting, since this cannot occur outside the head but need not be limited to the tongue).

Even if we ignore these findings and consider gustation as the perceptual modality of taste, it is questionable if we can even consider basic taste perception as representational. We do not have conscious access to decomposing the components of a taste experience (Stevenson, 2009a), which even on the pluralist conception of senses endorsed by Fulkerson (2014) might be a necessary condition. Additionally, taste considered just as gustatory sensations on the tongue will not satisfy enactivist theories of representation (Gray & Tanesini, 2010), and (if less embodied theories of perceptual representation are your thing) it is unlikely that we can generate an account of our perceptual representation of the concentration of basic tastes or introspectively access the gustatory components of the flavor experience (Lycan, 2018).[5]

If taste does not represent in a sense attributable to other senses, we might attempt to relax the criteria for it being a perceptual capacity by employing object perception criteria, such as constancy effects or figure–ground segregation.[6] And although consumer research on the bliss point for sweetness might provide a means of generating an account of perceptual constancies (for an introductory account, see Taubes, 2016), it is hard to conceive of what figure–ground segregation might amount to for basic taste qualities. However, my failure of imagination should not be taken as a reason to conclude it is not possible but merely improbable, thereby justifying exploring alternative and more likely lines of theory construction regarding taste perception. Based on the aforementioned philosophical reasons, empirical research, and interest in maintaining some aspects of our pre-theoretical conception of taste, it seems safe to conclude that taste is not merely gustation.

3.3 Expanding the Organ of Flavor Perception

The perceptual modality of taste should not be limited to just gustatory sensations, at least when just considering sensory qualities, as this limits the scope of what we ordinarily consider to be taste experiences. Pure gustatory sensations are not possible outside of laboratory conditions, and even when possible, most subjects report the pure sensations of basic tastes do not correspond to naturally occurring perceptions of these same qualities when combined with the other flavor sensory qualities. Lastly, it is questionable if taste perception conceived as gustation can even be considered representational.

The example of eating while suffering from nasal congestion (section 3.1) nicely illustrates that we commonly understand that gustation on its own is not responsible for our experience of taste perception. Upon learning that retronasal olfaction contributes to the experience of consuming food or drink, people acknowledge that retronasal olfaction plays a role.

Pragmatically, even in laboratory settings, it seems nearly impossible to isolate gustatory sensations independent of the other sensory systems within the oral cavity. While it is possible to generate a tastant without odorant qualities, attempting to stimulate the taste receptors without also activating tactile and thermal receptors is extremely difficult. In instances of pure tastants without odor qualities being placed on the tongue and controlling for somatosensation by having the stimuli placed within a flow of synthetic saliva at body temperature, subjects describe these experiences as odd and have difficulty identifying the gustatory qualities with the tastant's usual sensory qualities (Spence et al., 2015). Not only does this show the difficulty of isolating and presenting gustatory qualities on their own, but also, in a similar fashion to retronasal olfactory qualities, these are not identified by subjects as equivalent to those experienced in everyday multisensory taste experiences, making it reasonable that at least for olfactory sensory qualities, our sense of taste must usually encompass the perception of both gustatory and retronasal sensory qualities.

We can, of course, introspectively pretend to abstract taste experiences from the other sensory systems that provide peripheral and direct flavor experiences, but this would not reliably capture what we phenomenologically experience as taste. Introspecting our past experience, while a blunt theoretical tool, nicely demonstrates that taste is more accurately described as occurring within our mouth. Perceiving flavors requires multiple sensory pathways to encode fully the variegated properties of the object placed within our mouth. Yet, this multidimensional multi-chemically phased entity is perceived as a unified entity (Auvray & Spence, 2008; Stevenson, 2009a). We don't just taste with our tongues—we taste with our mouth. The sensory systems in our mouth allow us to sense food's tactile, thermal, and somatosensory qualities. But why stop here? Why not expand the organ of flavor perception to include all the sensory systems that transduce flavorful and sapid entities placed within our oral cavity?

What we consider to be taste might further be attributed to the volatile chemical subparts of the stimulus being pushed through the back of the nasopharynx and into contact with the olfactory receptors. Once it is noted that we should expand our notion of tasting to include tactile, thermal, and somatosensory sensation within the oral cavity, it seems quite reasonable to expand further what we consider to be taste as including retronasal olfaction, thereby requiring a refinement of our intuition that we taste things within our oral cavity to include the headspace of the odor plume as it traverses the nasopharynx and interacts with the olfactory epithelium.[7]

Gustation on its own might not be considered a perceptual modality, although it does generate individual sensory receptivity to particular types of chemical stimuli. We use our mouth, including the nasopharynx passage of retronasal olfaction, to perceive flavors. We taste flavorful and sapid entities. What we taste with our mouth are objects that exist across multiple chemical phases with a variety of different types of properties. For example, even the vapidest custard will have textural qualities, gustatory sensations of sweetness, and olfactory hints of vanilla. Taste requires the use of many sensory systems to encompass fully the variety of properties inherent within the object placed within the oral cavity. Nonetheless, it yields a unified percept of a flavorful object (Auvray & Spence, 2008; Spence, 2015; Stevenson, 2009a). A better understanding of taste requires expanding what is considered the sensory organ for individuating flavor experiences together with what might be consider the common sensible responsible for flavor perception. But what are flavors? And moving onto the second of our nested issues, the object of experience, what is the object of flavor perception?

3.4 The Object of Taste Perception and Its Flavorful Properties

Flavors are properties of consumable objects. While it might be wondered what this will rule out, intuitively we understand that we cannot taste wind, rainbows, or magnetic fields. Flavors encompass a set of properties far wider than those things we considered to be edible (or can fit in our mouth). Flavor properties include things that are possibly consumable but not considered foodstuffs, as can be observed by those suffering from pica.[8] As a folksier example, we can all recollect what metallic doorknobs taste like, but hopefully we have not licked one recently.[9]

The consumable object that we taste is a dynamic entity with a range of chemosensory properties. The solidity, density, viscosity, temperature, texture, gustatory qualities, and retronasal olfactory qualities of the object within our headspace all combine to yield our taste perception of flavor (Auvray & Spence, 2008; Spence, 2015; Stevenson, 2009a). Each of these sensory channels provides its own contribution to the unified perceptual experience (Prescott, 2015; Small, 2012), all while the perceptible object is dynamically evolving through palpitation, mastication, or swallowing and the perceptible object's own shifting chemical phases as the flavor profile evolves. Ice cream serves as a wonderful example. The object begins as a solid upon our tongue and slowly liquefies, surrendering its sweetness, the viscosity and density of the creamy fats, as well as the gaseous headspace containing olfactory qualities. Each of the components can be independently manipulated under natural tasting conditions, but the experience that presents itself is synthetically complex—that is, the content is complex, yet the individual components do not immediately present themselves without some further attentional or motor-sensory probing (Stevenson, 2009a). But it might be wondered if taste is a multisensory perceptual modality whose object of perception are multidimensional, multiphasic chemosensory consumable entities (placed within the oral headspace), what are its flavorful properties? Having expanded the organ of taste perception to encompass a host of sensory channels including retronasal olfaction, which thereby handles the nature of the object of perception within our headspace, we now turn to the nature of flavorful properties.

The most tempting view of flavors is that they are properties wholly dependent upon the subject and generated by the subject in perceiving the stimulus. Given that flavors are not apparent to our dominant visuocentric conception of reality, it is tempting to assume that these properties exist merely in the mind of the perceiver. However, B. C. Smith (2009) carefully argues that naive subjectivism is a mistake.[10] It is not that the epistemic access to flavor profiles is not dependent upon a perceiver, but rather what requires careful theoretical analysis is the role the perceiver plays in gaining access to these flavor properties. According to Smith, it would be a mistake to assume that flavors are wholly subjective, mind-dependent entities that can only be known through the individual's transparent and infallible grasp of what they claim to be the content of their perception. Rather, we must distinguish between what is being experienced and the experience thereof. What we perceive when we taste flavors is something objective about the very nature of the consumable object, thereby allowing us to distinguish between the flavorful properties of the object that we can all gain access to and the content of perception that requires attribution of an experience relative to a subject.[11] A further strength of his account is that flavors, considered as objective properties of a certain type of perceptible object, allow for the generation of correctness conditions for the range of perceptible qualities accessible within a flavorful object. Correctness conditions can then be utilized to assess descriptions of taste experiences as representationally accurate, a misperception, or inaccurate.

If the object of flavor perception is a multidimensional consumable entity that occurs across chemical phases, then it makes sense that in perceiving the full flavor profile, one must account for the viscosity and skillful movement of the liquid across the tongue's surface to generate the tactile, gustatory, and olfactory sensory qualities required to perceive the full range of flavorful properties within something like wine for example. The wine expert employs multiple means of manipulation to probe the vast number of flavor properties. Skillful manipulation of the liquid enhances access to the qualities by generating a more dynamic odorant headspace, which allows a diachronic experience of the full range of retronasal olfactory qualities within the wine. Additionally, the motor-sensory palpitation and probing of the wine on the tongue and throughout the mouth increase access to the tactile, chemothesis, and gustatory qualities of the flavorful object. Yet, it must be noted that the perception of flavor is not simply the additive combination of each of the separate sensory qualities from each of the sensory systems used to transduce the flavorful object. As noted above, both gustatory and retronasal sensory qualities are reported as perceptually different in their common everyday multisensory presentations by comparison to their experimental presentations when the other sensory qualities have been subtracted. Thus, the formative representational nature of flavorful perceptual experiences should also be considered in individuating taste as flavor.

3.5 Multisensory yet Unimodal Perception

A strong reason for considering a host of sensory systems as one perceptual modality is that they generate a unitary percept. The flavorful object is itself a dynamic entity with a range of sensory properties. Yet, our perception of it yields a unified complex experience. Each of the components can be independently manipulated under natural tasting conditions, but the experience that presents itself is synthetically complex—that is, the content is complex but the individual components do not immediately present themselves (Auvray & Spence, 2008; Spence, 2015; Stevenson, 2009a). Without skillful training employing learned conceptual categories or attentional modulation coupled with fine-grained sensorimotor skills to probe the flavor object, we experience flavors as unified experiences of the object placed in our mouths.

The variegated and highly dynamic nature of the object of perception nevertheless presents flavor experience as of a non-decomposable compositional percept the evidence for which, including the unified nature of flavor perception, is carefully summarized by Stevenson (2009a). Considering the percept at a set time (synchronically), we notice that the parts are not automatically and non-inferentially presented. Our experience of flavors formed from multiple sensory pathways presents itself not as a simple entity but rather as a unified complex synthesis of parts of which we cannot, in the given moment, decompose it into its component parts. Given the predominance of retronasal olfaction for flavor, this is unsurprising, as olfaction generally implements a functionally compositional system of representations that is often not decomposable into its component parts (Young, 2015, 2019a; chapter 4).[12]

There are situations where we can manipulate the different components within and between sensory streams to highlight components of a flavor percept (e.g., menthol; B. C. Smith, 2015), but these are outliers and should not form the sole basis for claiming that a sense is multimodal. Examples of unified perceptual modalities that allow for experiences of components can easily be generated. Yet, we do not employ these as counterexamples for other modalities, such as vision being unimodal. For instance, when appreciating a cubist painting, the observer is able to see the use of edges that generate forms independent of object boundaries, or when viewing a pointillist piece of art, we can visually perceive color independent of shape. Additionally, we can use a ganzfeld or immobilize the eyes to demonstrate the necessity for motor-sensory input for visual perception, but from these examples alone, we do not then generalize that visual experience is multimodal.

Often, claims proclaiming our access to components of the flavor experience derive from ongoing perceptual states with attentional modulation. Serially probing the perceptible object reveals more of its features, but this does not negate the synchronic experience of unified flavor experiences. There is some evidence that, in applying conceptual categories together with practice applying these resources to sensory qualities of the perceptual class, wine experts show an advantage when compared to even proficient nonexpert tasters in recognizing the component sensory qualities of wines (reviewed in Stevenson, 2009a).[13] However, in the vast majority of cases, individuals—and even experts—have great difficulty perceiving the component parts of a multisensory flavor percept (Shepherd, 2012). Thus, even in these outlier cases involving what might be consider learned capacities, the format of the flavor percept does not reveal its compositional parts in a transparent fashion. Combining this conclusion with the those of the previous sections, we end up with the claim that taste includes activity from multiple types of sensory channels located throughout our headspace that transduces a dynamic chemical stimulus into a unified synchronically non-decomposable percept.

3.6 Conclusion

What has been offered throughout this chapter is a sustained argument that a comprehensive explanatory program can be used across the chemosensory systems that has been shown in chapter 2 to work for orthonasal olfaction as our perceptual modality for smell (Young, 2016, 2019a). According to this explanatory strategy, a chemosense can be individuated by determining its sensory (distal) object of perception, its unique sensory qualities, and the intentional object of experience/perception (including its representational format). Considered in this manner, it has been argued that we taste flavors.

Our perceptual experience of taste is not merely basic tastes on the tongue, nor is the object of perception a static entity. Rather, what we experience and represent as tastes are flavorful entities with multiphasic facets that require different sensory pathways to encode the range of properties fully that generate the unified experience of flavor qualities. The objects of taste perception are those entities placed within the oral cavity that our sensory transducers can access through their independent means of encoding the sensory qualities relative to their receptors range of sensitivities. Conceiving of the modality of taste as perceiving flavors allows material objects across multiple phases to be objects of taste perception. In the most simplistic of situations, our experience of flavor is composed of gustatory, textural, thermal, somatosensory, and retronasal olfactory qualities (Auvray & Spence, 2008; Prescott, 2015; Small, 2012). Without the confluence of a range of these independent qualities, our taste of flavors does not yield the sensory experiences that we would recognize as taste perception.

Each of the component sensory systems of taste allows for access to different qualities of the object of perception. When mouth breathing (in the absence of flavorful objects), we can perceive the temperature of the airflow, the humidity level of the environment, the viscosity of the saliva coating our tongue, the position of the tongue, and so forth. Each of these senses allows for independent exteroceptive and interoceptive experiences of different experiential properties instantiated within the oral cavity. What is unique to flavor perception is the addition of gustatory and retronasal stimulation (either in combination or, at times, on their own). When we have an experience of flavor, the percept is as of a unified object that cannot be decomposed within synchronic experience into its independent sensory components, at least in the case of retronasal and gustatory qualities (somatosensory, thermal, and chemothesis properties are to some extent dissociable; Stevenson, 2009a). What unifies our taste of flavors is the non-decompositional format of the combinatorial representation of the sensory qualities of the multiphasic chemical entity placed within the mouth, including its odorous headspace. Flavors might be objective properties of environmental entities, but we do not perceive them outside of our heads.

Taste is the perceptual modality that allows us to perceive multidimensional (phasic) chemical objects, with each aspect being transduced by a separate sensory channel that yields a unified synthetic experience in which the component's sensory qualities are not synchronically available for conscious experience. The unified format of flavor is thereby similar to that of smell in generating functionally compositional representations that withhold access to their complex internal structure, which will form the focus of the next chapter.[14]

4 Formatting Odors

Despite our preconception that smells do not occupy much real estate within our cognitive economy, it turns out that, with a little nudge, we readily admit to the significant role olfaction plays within our everyday lives. Not only do we smell complex odors throughout the day and, at times, notice the continual flux of the smellscapes we inhabit, but we also think about olfactory objects, make behavioral plans to engage with these entities (especially flavorful entities), and even converse with each other about smells and smellscapes. Given that our perceptual and cognitive states are about smells as distal entities as well as mental particulars (i.e., intentional objects within perceptual and cognitive states), it seems arguably the case that even if olfaction is excellent at detecting odorants and changes within the chemical environment (Barwich, 2014; Keller, 2017), it is capable of much more, such as perceiving smellscapes (Young, 2020) and amodal perceptual object completion (Young & Nanay, 2021).

This chapter focuses upon the representational format of olfactory experiences. To do so, it will first establish that our smell experiences can be representational and then transition to providing an account of the kind of representational system implemented by smell. With this remit, the further issues of cross-cultural linguistic mediation of our olfactory ability to categorize smells and how we can cognitively ponder smells (categorically) will be left aside until the next chapter.

Analyzing the representational format employed by the olfactory system has the added value of yielding an understanding of nonconceptual content as nonconcatenative, compositionally formatted mental states. To generate this conclusion, the representational status of olfaction must first be established. Once it is allowed that olfactory states are representational, the nature of their representational format is identified in terms of the type of compositionality implemented within the olfactory system—and not just for smells. The representational format of olfaction is also found within retronasal olfactory systems. Despite using the nested questions of what smells are (chapter 2) to distinguish smell from retronasal olfactory flavor perception (chapter 3), it turns out that olfactory retronasal flavor perception is similar in terms of its synthetic percepts. Thus, this chapter shows how olfaction provides a novel means of reinvigorating the debate about nonconceptual content and an empirically viable example of how a neurobiological system could implement a nonconceptual representational system, while also explaining the puzzling discrepancy between our gifted ability to detect and discriminate odors and our meager capacity to identify smells.

4.1 Smell Experiences Are Representational

To establish that our smell experiences have representational content, a range of criteria for the representational status of the content of mental states and perception will be applied. Rather than taking one standard of representational content as the sole criterion, the methodology employed will survey multiple plausible criteria. For instance, if it can be shown that our olfactory experiences do not simply causally covary in a manner that suggests a mere isomorphic constant conjunction, then the content of our olfactory experiences are representational. Or alternatively, if our olfactory cognitive states occur in a manner that is not simply attributed to a closed causal system of stimuli to sensory state pairings, then I will presume that our experiences of smells have representational content. The key tests of the claim that olfactory states are representational will be whether misrepresentation, intentional inexistence, and representing the objects of experience according to different modes of presentation can occur within olfactory experiences.[1]

One of the key tests for representational content that separates perceptual states from mere sensory states is whether the state is brought about by isomorphic causal covariation, such that if a given stimulus is in the receptive field of the animal, it automatically creates a behavioral or sensory response (Dretske, 1981; Fodor, 1987, 1990; Fodor & Pylyshyn, 1988; Millikan, 1984, 1993, 1994, 2000; Papineau, 1987, 1993). The stimulus–response behavior of cognitive states varying between stimulus detection and transduction should determine if the content of the state is representational in any sense. Thus, our starting point in demonstrating that olfactory perceptual states can be representational will be showing that we can misrepresent what we smell in terms of both misattributions of the causal source and the misidentification of the object of perception.

Similarly, it was because of concerns regarding the causal isomorphism between stimulus and sensational response of olfactory experience that led sensationalists such as Reid (1764/1997) to argue that olfactory experiences could not be representational. According to Reid,[2] since we are forced to undergo an olfactory experience whenever the causal source is present and we can only have the experience when the causal source is present, then the olfactory experience must be sensational. However, if it can be shown that we may abstain from undergoing olfactory experiences, even in the presence of an olfactory stimulus, or that we can elicit olfactory experiences in the absence of a causal stimulus in a manner that is not explicated purely in terms of brute physiological processes, then this should be sufficient to show that these states are representational. Thus, examples for olfactory experiences satisfying intentional inexistence, including hallucinations, olfactory imagery, and amodal completion, assuage this further line of worry about the representational status of olfaction.[3]

4.1.1 Misrepresentation

A simple test of whether a given state is representational is if it can misrepresent. If the content of the state can vary across presentations of the same stimulus, then the olfactory system must be capable of representing the initial sensory state in different ways that go beyond mere causal connections. Olfactory misrepresentation most commonly occurs in two different forms: misattribution of the olfactory experience's causal source and misidentification of the olfactory object of perception.

Misattributing the causal source of an olfactory experience occurs more frequently than one would expect. In these cases, we do not misrepresent the olfactory quality of the odor, but we misrepresent what perceptual object we think we are smelling. The most prevalent contemporary cause of misattribution can be attributed to industrially produced synthetic compounds with odor profiles almost identical to naturally occurring smells. For example, when walking by a sandwich shop with its sumptuous smell of freshly baked bread, sometimes the causal source of your olfactory experience is not freshly baked bread but rather odorants disbursed via aerosol cans. There is a sense in which the experience is not misrepresented because the olfactory quality generating what we identify as the smell of freshly baked bread might be realized by different complexes of molecular compounds that are present in the vicinity of both freshly baked bread and aerosol cans. The olfactory quality produced by both possible causal sources is qualitatively almost identical. However, in this instance, the attribution of the smell gets the causal source wrong. While misattributing the causal source of our olfactory experiences is not a gross misrepresentation, given that the synthetic chemicals are engineered to produce a very similar type of olfactory quality, the exact chemical structures transduced, particularity of each token experience, is nonetheless different. To unpack this example, you are undergoing an experience that presents itself accurately as the smell of baked bread (i.e., your experience of the olfactory quality of baked bread is veridical), but you are not smelling freshly baked bread as the causal source of the experience.

Another example of misrepresenting the content of our olfactory experience is misidentifying the object you are smelling. For instance, human vomit and Parmesan cheese share some key chemical compounds responsible for their distinctive olfactory quality (Gilbert, 2008). While the odorants composing each of these olfactory entities are not chemically identical and vomit and Parmesan smell different, nevertheless, under ambiguous conditions, subjects presented with one compound commonly mistake it for the other and accordingly rate their valence and hedonic value based on the perceived identity of the object. From my own experience, this example works best when you present audiences with an airline vomit bag filled with fresh grated Parmigiano Reggiano at room temperature and tell them you were traveling with a baby. Thus, we misrepresent the identity of the thing we are smelling, including its properties of valence and hedonic value.

Misattribution and misidentification taken together support the representational status of our olfactory experiences. Based on these examples of the misrepresenting, we can infer that the content of our olfactory experience occurs in some representational medium. Consequently, the constant conjunction of stimulus and response in a manner that does not allow for causal variation across pairings of stimulus and mental state is ruled out by these examples.

4.1.2 Intentional Inexistence

A further criterion for representational content is whether the state can be contentful in the absence of a causal stimulation. Can the state represent something in the absence of a perceptual object, or can it represent something nonexistent? If our olfactory experiences can occur in the absence of an olfactory stimulus, then this will at least refute Reid's second claim that we cannot have an olfactory experience in the absence of an olfactory object. Below are a non-exhaustive bunch of examples of contentful olfactory experiences in the absence of a perceptible object, covering disorders of the olfactory system, hypnosis, dream states, and mental imagery.

4.1.2.1 Smell disorders Olfactory hallucinations occur across human development in a variety of types, comorbid with other disorders and on their own, as well as in similar non-veridical states such as dreams and during hypnosis. Experiences of olfactory hallucinations are often described in terms of the perception of an unpleasant olfactory quality. Some experiences include a directional component. Yet, most olfactory hallucinations are not described as occurring in egocentric space. Surprisingly, some cases have been reported in which the hallucinatory percept occurs in just one nostril, which is at vast odds with our everyday perception of smells. Olfactory hallucinations are similar to perception in terms of the types of experiences reported, cortical areas activated during hallucinations, and dependence on olfactory memory. This section covers dysosmia, a perceptual disorder that borders on hallucinatory; phantosmia, where subjects report the experience of phantom smells; and olfactory dreams and hypnotic experiences.

Dysosmia, also referred to as troposmia or parosmia, is the pathological condition in which individuals correctly identify their experience as being of an olfactory quality, yet their reports of the olfactory quality incorrectly describe aspects of the quality (Leopold, 2002). In some cases of parosmia, the individual reports that the experienced olfactory quality is unusual and not fully describable. Given that instances of this condition are generated by stimulation of the olfactory receptors, they are more aptly conceptualized as cases of misperception.

Phantosmia, the olfactory experiences of phantom smells, are classic cases of hallucinatory experiences. They are characterized in terms of the occurrence of an olfactory quality in the absence of sensory stimulation. Some instances of phantosmia are caused by sensorimotor activity such as sniffing, laughing, coughing, or exhaling (Leopold, 2002), but they also occur independently. Olfactory hallucinations are often comorbid with other disorders such as depression, schizophrenia (Arguedas et al., 2012a, 2012b), and epilepsy (Henkin et al., 2013a, 2013b) and occur throughout the entire population across age groups (Sjölund et al., 2017).

Based on subjective reports, the hallucinatory experience is often of negative olfactory qualities (cacosmia is a good example, which smells exactly as it sounds) or often described as having a smoky or burning odor quality. In some rare cases, subjects include descriptions of more pleasant experiences such as smelling flowers (reviewed in Sjölund et al., 2017). The majority of experiences of phantosmia are reported as having an olfactory quality that is low to moderate in its intensity and vividness, which is in keeping with reports of olfactory imagery that often report the experienced quality as less intense or vivid than those qualities of other perceptual modalities (Dou et al., 2018). Olfactory hallucinations cannot be attributed to deficits in olfactory identification (Arguedas et al., 2012a, 2012b) but are more likely linked to olfactory perceptual processing in the cortex, making them good examples of olfactory perceptual representations of an odor as being present in terms of the experienced olfactory quality, even in the absence of direct sensory stimulation.

4.1.2.2 Hypnosis Another example of olfactory experiences with a qualitative component of smell in the absence of direct sensory stimulation might be derived from the few instances of olfactory hypnosis. Studies of olfactory experiences under hypnosis are rare, but Barabasz and Gregson (1979) document experiences of odor experiences induced by hypnosis that yielded the same measurements of galvanic skin response as perceiving the smell, suggesting a similarity of experience of olfactory quality using this secondary measure.[4] Similarly, Cox and Langdon (2016) were able to induce both positive and negative olfactory experiences hypnotically, such that when the odor percept was induced, it was claimed as intense, while negative hypnosis decreased subjective reports of identifying the target olfactory quality and its intensity.

4.1.2.3 Olfactory dreams Another example showing that olfactory experiences can be representational in the absence of olfactory stimulation is olfactory dreams. The occurrence of dream experiences with olfactory qualities was documented by Stevenson and Case (2005). Using self-reports of dream experiences, they discovered a strong relationship between individuals who experience olfactory qualities during dreams and the ability to generate olfactory imagery volitionally. Reports of olfactory dreams are infrequent and in keeping with the general findings about olfactory imagery and hallucinations in terms of both frequency and lack of vividness.

Olfactory hallucinations, hypnotic odor experiences, and olfactory dreams are all identified in terms of their olfactory qualities mimicking perceptual states. These phenomena cannot be attributed to deficits in olfactory identification, the motivation of the experimental participants, or attention. Rather, they are likely connected to the role of olfactory memory that predominantly encodes and stores smell experiences with a negative valence (Larsson et al., 2009; Stevenson & Case, 2005). Given the central role of memory, as well as the reported experiences as representing odors without direct stimulation by an odorant, these examples show that olfactory experiences can occur in the absence of an olfactory stimulus, which suggests that our olfactory experiences have representational content.

4.1.2.4 Olfactory imagery Olfactory hallucinations are not the only instance of olfactory experiences occurring in the absence of a stimulus, as there is a large and growing experimental literature on olfactory imagery (reviewed in Arshamian & Larson, 2014; Young, 2019c). Olfactory imagery is the occurrence of an experience with an olfactory quality component in the absence of sensory stimulation at the receptors by an odorant. Olfactory imagery demonstrates that we can experience smells in the absence of an odor stimulus. A good example of volitional olfactory imagery is derived from an anecdote told about a professor who, on a particular day of class, announced while opening an empty utility jar that he had released an unpleasant odor in the lecture hall and asked that people raise their hand when the odor reached them. Even though no odorant was in fact released, hands went up in a manner that simulated an odor plume moving backward toward the top of the lecture hall. Taken on its own, this might not suggest that they were undergoing a smell experience but might be interpreted as the class tacitly employing their knowledge of chemical diffusion. However, some of his students claimed the odor they were experiencing was so noxious that they needed to leave to deal with their nausea. The story suggests that people can have olfactory experiences in the absence of any stimuli that even mimic the ability of noxious odors to induce nausea.

(A more detailed introduction and analysis of olfactory imagery can be found in Young. 2019c.)

4.1.3 Representing *x* as *f*

Another test for being a representational system might be whether the stimulus can be represented in different ways. Can we accurately identify or report the olfactory object yet, in so doing, perceive it in different ways, given its properties and our associative categorical repertoire for smells? Do olfactory modes of presentation exist?

Examples of representing a smell as instantiating certain properties or under different guises can be derived from perfume. Perfumes can be identified as being of a given brand, such that one can smell that a perfume was created by a particular nose, that it is some manner of a given product range (e.g., flankers), and quite often, some brands will have a distinctive quality that makes them recognizable as belonging to the same company. Additionally, perfumes as a form of fashions have trends. The smell of perfume *x* might be recognized as being of the latest fashion. These perfume trends make it possible to identify a scent as being from a particular time period. Our ability to represent an olfactory object under multiple and dissociable guises demonstrates that our olfactory experiences have modes of presentations, thereby providing evidence that our experience of smell occurs in some representational medium. The examples surveyed in this section of misrepresentation, intentional inexistence, and representing olfactory experiences as *x* all rule out that the relation between the olfactory object and our experiential states as simply a matter of isomorphic causal connection. The contents of our olfactory experiences are representational.[5]

4.2 The Nonconceptual Format of Olfaction

Having established that olfactory experiences are representational, the next stage in our theoretical progression of understanding the nature of smell is to tackle the further question of what the format of representation employed by the olfactory system is. And it turns out that a novel understanding of nonconceptual content emerges from studying olfaction (Young, 2015). The olfactory system processes smell in a structural manner that is dissimilar from the common compositional system responsible for thought or language. Thus, our olfactory abilities will be used in this section to generate support for the theory of formative nonconceptual content.

The difference between conceptual and nonconceptual content at its most basic level is nothing more than distinguishing between different kinds of states (sensory, perceptual, or cognitive) based on their representational structure. Nonconceptual content should be defined in terms of the structural format of sensory, perceptual, or cognitive states ostensively defined by contrast with the structure of concepts. To arrive at the structural nature of FNCC, I suggest employing a contrastive methodology together with the process of elimination. Traditionally, the debate regarding nonconceptual content has progressed by identifying the criteria of concepthood and then contrastively demarcating other states that are not conceptual, given these conditions. In keeping with this methodology, I suggest sidestepping any theoretical confusion regarding concepts by using the most minimal yet widely agreed-upon conditions for being a concept to identify nonconceptual content by process of elimination as any sensory, perceptual, or cognitive representational state that does not obey these minimal conditions (Young, 2015).

Minimally, concepts are mental particulars, such that they can either be the constituents of thoughts or are employed in determining the content of cognitive states. Additionally, concepts are compositionally structured in some fashion, such that they can be employed across complex cognitive states and within inferential reasoning. Methodologically, the structural nature of concepts is inferred from a perceptual, linguistic, or cognitive capacity in concert with the structure of the target domain. Traditionally, this has led to the claim that concepts must be compositionally structured or, at the very least, sensitive to the compositional structure of the target domain. Thus, by process of elimination, any contentful state whose structure, as ascertained by our capacities and target domain, is not the same as the structure of the conceptual domain should be considered nonconceptual. By first looking at the compositional format of concepts, we can then compare this representational format to that employed by olfaction in support of the claim that olfaction employs a nonconceptual format of mental representation.

4.2.1 Concepts and Compositionality

Despite much disagreement in both philosophy and cognitive science regarding the nature of concepts, there is consensus that concepts are, to some extent, compositional (Connell & Lynott, 2014; Del Pinal, 2015; Frixione & Lieto, 2012; Jylkka, 2011; Piantadosi et al., 2016). Theories of concepts vary widely on the details of how our conceptual system implements and respects the notion of compositionality. Despite these differences, it is traditionally assumed that one of the essential characteristics of concepts is that they are minimally sensitive to and have access to their constituents.

There is a long-standing tradition of inferring the representational structure of our cognitive states from our cognitive or perceptual capacities. Most famously, this move is seen in the systematicity argument, where the compositional syntax of thought is inferred from the compositionality of language (Fodor, 1987; Fodor & Pylyshyn, 1988; Fodor & McLaughlin, 1990). However, this strategy is also found in the productivity argument (Fodor, 1981, 1985, 1987; Fodor & Pylyshyn, 1988), visual and auditory systematicity (Cummins et al., 2001), the tracking argument (Horgan & Tienson, 1994, 1996), and our capacity for inferential coherence (Crane, 1990; Fodor, 1985; Horgan & Tienson, 1994, 1996). Our mental economy's systematic and productive nature, together with the stability of conceptual content, intrasubjectively across time and intersubjectively at any instance, requires that our conceptual systems must be compositional. In what follows, I shall remain neutral regarding the nature of concepts and only assume the minimal consensus for the sake of argument that whatever concepts might be, they must obey compositionality to some extent. (Young, 2015).

Compositionality is determined relative to a system of representations, such that it requires specifying the primitives for the system, the rules of formation and transformation, as well as the rules of well formedness allowed by the system for generating meaningful complex expressions (van Gelder, 1990). The minimal conception of compositionality requires a specification of the interrelation between the syntax and semantics of a system of representation, such that the meaning of a complex expression is determined by the meanings of the atomistic parts and the system's mode of combination (Dever, 2006, 2012; Pagin and Westerståhl, 2010a, 2010b). The most important factor in determining whether a system is compositional is the mode of combination—that is, the rules governing the composition of complex expressions. The mode of combination determines whether a system is compositional, as well as the type of compositionality.

4.2.2 Concepts and Their Relation to Classical Concatenative Compositionality

The most stringent form of compositionality occurs in Fodor's (1976, 1981, 1987, 2000, 2008) language of thought hypothesis (LoTH), according to which thought occurs in a representational medium similar to language, such that each thought is the concatenation of its constituents. According to the LoTH, all thoughts must occur in such a compositional rubric as evidenced by the systematic and productive nature of our cognitive and linguistic abilities (Fodor, 1981, 1987; Fodor & Pylyshyn, 1988; Fodor & McLaughlin, 1990). Classical compositionality is concatenation based on the spatial juxtaposition of expressions within the complex mental representation. The internal structure of the system's states must be of a formal nature, such that the constituency relations among the expressions are directly mirrored in the structure of the corresponding tokens. Thus, the explicit syntactic representation of the primitives within the complex mental representation is required.

With the rise of research on our psychological conceptual capacities, their inferential roles, and their biological realization, it has become clear that while concepts require compositionality, how they implement it varies from the logical part–whole concatenation espoused by the LoTH. Contemporary views of concepts still hold compositionality dear. Yet, their means of implementation can vary drastically.

Prototype theories of concepts stand in stark contrast to the atomistic theories of concepts predicated upon a concatenative internal syntactic structure. According to prototype theory (and most statistical regularity theories of concepts), the complex structure of a concept encodes a statistical typicality relation of the properties their members tend to have (Margolis & Laurence, 1999, p. 27; G. L. Murphy, 2002). Recognizing the challenge of generating compositionality within these systems (Fodor, 1981; Fodor & Lepore, 1992), prototype theories have offered various implementation strategies for accomplishing classical compositionality (Del Pinal, 2015; Frixione & Lieto, 2012; Prinz, 2002, 2012).

However, more scientifically plausible and biologically implementable theories have been offered that create compositional concepts not in terms of concatenative representations. These transitional theories still maintain that complex concepts require constituent structures to be compositional. However, they weaken the requirements for generating constituent structures, either by employing a multivariate system that respects classical compositionality using biologically plausible versions of tensor product networks (Stewart & Eliasmith, 2012) or by allowing that the constituents do not need to be literal components of the complex representation (Rice, 2013). What unites the aforementioned theories of concepts is their adherence to the necessary condition that compositional representations require internal constituents or need to be connected via composition or decomposition functions to the relevant constitutive concepts.

4.2.3 Functional Compositionality

Minimally, all that is required for a system of representations to be considered compositional is the specification of a relation for the combination of expression types and primitives. Recognizing the theoretical possibility of a more minimal form of compositionality, van Gelder (1990) proposed functional compositionality, which requires a systematic manner for creating compound expressions given the constituents and a similar systematic mode for decomposing the complex representations into their constituents. Mental representations with functional constituent structures are composed from the constituents and can be decomposed back into them, but these complex representations do not have constituent parts that we can refer to as their internal structure.

van Gelder's (1990) theoretical motivation for formulating compositional representations without discrete syntactic constituents was to explain how connectionist networks could generate systematicity. With systematicity as his target phenomenon, van Gelder proposed functionally compositional systems as requiring reliable rules of processing for generating a complex from the constituents, decomposing these functionally complex representations, and transforming these same constituents into systematic variants (Martinez-Manrique, 2014, p. 309). For van Gelder's purposes, the decomposition function plays an integral role and is a necessary condition for a representational system to be functionally compositional (van Gelder, 1990, p. 361).

However, a more minimal form of compositionality is possible that jettisons the ability of the system to decompose the complex representation functionally into its constituents. The most minimal form of compositionality requires not that the components persist within the complex expression or that the complex be decomposable into its previous component parts, but only that the meaning of the whole is determined from the meaning of these parts. The functionally complex representation is endowed with complex meaning within the system by virtue of being composed from constituents. Thus, it is a theoretical possibility that a complex representation with constituent structure need not require an explicit internal syntactic structure of the classical type, nor that it be decomposable back into these constituents. The minimal conception of compositionality allows for a permissive stance on compositionality that requires only that the complex representation be functionally composed from the constituents. Our olfactory sensory, perceptual, and cognitive states can be used to establish that the olfactory system employs this minimal non-decomposable form of functional compositionality.

4.3 Formative Nonconceptual Olfactory Content

Olfactory sensory, perceptual, and mental imagery states occur in a functionally compositional format, such that their constituent structure does not obey the strictures of concatenative compositionality that is required to be considered a conceptual state. Using the contrastive methodology, these results suggest that the difference between conceptual and nonconceptual content at its most basic level might simply be a distinction between different kinds of states based on their representational structure.

FNCC, as I have dubbed it (Young, 2015), is nothing over and above the thesis that nonconceptual content should be defined in terms of the structural format of states ostensively defined by contrast with the structure of concepts. The difference between conceptual and nonconceptual states is simply a matter of the format of their structural parts and relations within a system of representations. FNCC is offered as a minimal and precise conception, which it is hoped can be utilized to clarify the nature of nonconceptual content.

In keeping with representational pluralism (Cummins, 1996; Cummins et al., 2001) and nonconceptual pluralism (Bermudez, 2007), what is on offer is a general methodology and inclusive definition of nonconceptual content. The functionally compositional format employed by the olfactory system is offered as a sufficient condition for nonconceptual content and should not be thought of as an exhaustive account of all types of nonconceptual content. Furthermore, the pervasiveness of formatively nonconceptual kinds should not be considered theoretically perverse. The number and kinds of nonconceptual states are not up for theoretical speculation, as it depends upon empirical research. Every cognitive or perceptual ability that is empirically shown to be representational but not conceptually formatted should be considered formatively nonconceptual.

4.3.1 Functionally Compositional Olfactory Representations

Olfactory processing does not always employ concatenative compositionality. Olfactory states are formatted in a manner that is mediated more by sensory templates than by our conceptual repertoire and lexical resources (Young et al., 2014). This section explores the compositional format that mediates olfactory encoding of odorants and proceeds by extrapolating the compositional structure of olfactory states from olfactory sensory transduction, perceptual encoding, and mental imagery.[6]

4.3.1.1 Stimuli transduction and syntactic encoding The most striking difference between olfaction and the other sensory systems is the lack of isomorphism between receptor types and perceptual stimuli. There is no strict chemotopic mapping of chemical properties of the distal stimuli to individual receptor types or higher-level glomeruli in the olfactory bulb. The lack of chemotopic maps either at the receptors or OB is quite unlike vision with its retinotopic maps and orientation columns in V1, or audition with the decomposition of frequency in the cochlea. Furthermore, olfactory cortical states encode complex odor compounds not simply as the summation of their constituents, which begins to provide the basis for an explanation of our inability to identify more than four components of complex olfactory mixtures perceptually (section 4.3.1.5), which arguably shows that olfactory perceptual experiences are functionally compositional in the weak non-decomposable sense.

4.3.1.2 Olfactory receptor neurons and the olfactory bulb Odorants are transduced in a combinatorial manner (Araneda et al., 2000; Buck & Axel, 1991; Firestein, 2001; Hallem & Carlson, 2006; Hallem et al., 2006; Malnic et al., 1999; Meierhenrich et al., 2004). At the initial sensory level, the olfactory system encodes the molecular structure of chemical stimuli in a distributed fashion across multiple olfactory receptor neurons. Odorants are combinatorially encoded in a spatially extended and parallel fashion across multiple receptors both at the receptor level and at the further stage of processing within the OB. The glomeruli and mitral cells within the OB encode input from across multiple ORNs. Glomeruli show a preferential firing pattern for particular chemical structures but are also sensitive to other chemical stimuli. As a result, a chemotopic map does not arise within the glomeruli of the OB. The OB might not have a strict chemotopic map like the retinotopic map of vision, but it does have a coarse chemotopic organization (Bozza & Mombaerts, 2001).

The combinatorial encoding of a stimulus is distributed across multiple regions throughout the glomeruli of the OB, with each glomerulus being sensitive to different parts and combinations of the chemical structure. Odorant encoding occurs in a parallel and distributed manner across multiple glomeruli. The distributed nature of olfactory stimulus encoding within the OB suggests that monomolecular and, by extension, complex odors are not encoded as the sum of their parts. Rather, they have their own unique distributed patterns of activation (for a review, see Auffarth, 2013). In addition to the spatially distributed encoding of olfactory stimuli across glomeruli, the encoding dynamic is also temporally extended (Haddad et al., 2013; Linster & Cleland, 2013; Olofsson, 2014). Olfactory stimulus encoding is accomplished by large groups of glomeruli and mitral cells firing across time, such that stimuli transduction is spatially and temporally distributed.

The distributed nature of the spatiotemporal structure of sensory representations might be used to adjudicate against concatenative representations with their strict internal syntax. Because the primitives within a concatenative compositional system are literal parts tokened within the complex expression, it is prima facie not obvious how this could be accomplished by the olfactory bulb's encoding of odorants. Literally tokening the constituents is not possible at a given time, since complex olfactory mixtures are encoded in a spatiotemporally parallel and distributed fashion within the OB. However, relaxing the temporal syntactic requirements of concatenative compositionality to allow for distributed temporal parts would still not circumvent the problem, as they would still be spatially distributed in a manner unlike any example that has formal parts within a complex representation. Hence, olfactory stimuli at the sensory level of olfactory processing are not encoded by a concatenative syntactic system.

Sensory transduction and olfactory stimuli encoding provide strong evidence that olfaction implements a syntactic system employing functional compositionality. What needs to be further demonstrated is that odorant encoding forms complex representations that generate compositionality that does not allow access to the constituents or that cannot be decomposed.

4.3.1.4 Odorant encoding at the piriform cortex The rough chemotopic organization found within the OB is not maintained in the olfactory system's projections to the piriform cortex.[7] In animal studies of the neural encoding of odorants, it has been shown that the PC does not maintain the chemotopic organization found in the OB. Rather, convergent neural ensembles respond to dissimilar molecular structures of odorants in the same way (Stettler & Axel, 2009). This is in marked contrast to the visual, auditory, and somatosensory systems that preserve the spatially organized transduction of the similarity of stimuli features from the sense organ through their respective sensory cortexes (Marshall et al., 1941; Talbot & Marshall, 1941; Woolsey & Walzl, 1942). Moreover, odorant encoding within the PC is represented in a sparse and spatially distributed fashion that does not code the minute differences between the structural components within complex odor mixtures (Illig & Haberly, 2003; Rennaker et al., 2007; Stettler & Axel, 2009; D. A. Wilson & Stevenson, 2006).[8] Additionally, recent research has progressed our understanding of the PC as an associative center for olfactory processing and as integral for generating configural odor objects from the molecular properties encoded at lower levels of sensory processing (for a review, see Courtiol & Wilson, 2016b).

While chemotopic organization is not maintained in the PC, it has been documented that cortical neurons within the dorsoposterior part of the anterior PC (APC) in rats display neural specificity to some food-related categories of odorants either in a holistic fashion to the entire mixture of odorants derived from the food source or selectively to a prominent component of the mixture (Yoshida & Mori, 2007). Despite cortical neurons within the PC displaying some manner of sensitivity to the odorant category, further research substantiates interpreting the PC as an associative network for generating categorical stereotypes of odorants not in terms of their concatenative compositionality but rather in a holistic fashion (Stettler & Axel, 2009). Barnes et al. (2008) found that neuron ensembles within the rat PC treat an odorant composed of ten components as equivalent to that same mixture minus one component. Their results indicate that the PC completes the representation of the complex odorant, even in the absence of a component. Additionally, their research showed the neural ensembles discriminated between two different ten-component mixtures. Together, these studies suggest that the olfactory system at the PC encodes odorant categories of varying molecular complexity in a combinatorial manner that does not depend upon explicitly representing each of the component parts. The combinatorial syntax of olfactory stimulus transduction within the PC does not depend upon the explicit representation of each constituent in a manner required for concatenative compositionality. Thus, it is arguably the case that olfactory processing at the PC implements a system of functional compositionality that does not require access to all of the constituents or the decomposition of the complex mixture into their constituents.

4.3.1.5 Olfactory mixture perception Further evidence that the olfactory system and, in particular, its perceptual states implement non-decomposable functional compositionally can be derived from research on olfactory mixtures. Olfactory mixtures occur when two or more odorants are combined to form a complex odor. The majority of smells are composed of a vast number of molecular compounds. Consequently, the nature of olfactory mixtures is of great importance in attempting to understand how it is that we recognize, identify, and individuate a smell. There is evidence for distinct encoding at the cortical level between single and binary odor mixtures (Boyle et al., 2009). However, in what follows, the focus will be upon complex odor mixtures, as these are the most decisive in examining the compositionality employed by the olfactory system.

When two or more odorants are combined into a complex, one of two possible mixtures results: a configural mixture whose olfactory quality differs from the smell of the components, which are not discernable as constituents of the new smell (e.g., the lemon and lime combination in 7 Up); or an elemental mixture whose olfactory quality is merely the concatenation of the components that are discernable within the complex (e.g., citrus bleach; see chapter 2 for a fuller introduction of olfactory mixtures). Configural mixtures are particularly fascinating, since the mixture's quality is not determined as an additive process, such that one can predict the new smell from its individual components (Berglund et al., 1973). What becomes clear is that the olfactory system can represent complex olfactory stimuli either in a classical compositional manner (elemental mixtures) or by employing functional compositionality (configural mixtures).

The combinatorial syntax of transduction at the sensory level sheds light on the next olfactory phenomenon of our inability to identify the components of olfactory mixtures, which suggests that smell employs non-decomposable functional compositionality. Livermore and Laing (1996, 1998), Laing and Francis (1989), Laing and Glemarec (1992), Laing and Jinks (1999), and Laing et al. (2001, 2002) have all documented our inability to identify odorants within a complex odor. They established that even if the subject had access to the individual constituent of the complex odor, they could identify at most three to four of the components within a complex chemical mixture. The inability to identify the constituents within a complex smell perceptually is best explained in light of the aforementioned evidence that the nature of sensory and cortical encoding of olfactory stimuli does not always encode complex odors as the concatenation of their constituents. The limitation in identifying parts of a complex odor might be attributed to the compositional format outputted from the PC, but even taken at face value, our perceptual inability to identify more than four components of a complex olfactory mixture suggests that all of the constituents of the mixtures are not perceptually accessible. Since we can accurately estimate the concentration of the constituents and we experience the mixture as a complex entity that transcends the summation of its constituents, it cannot be objected that these are simply cases of the olfactory system generating new primitives.

Configural mixtures confirm that our olfactory perceptual abilities do not represent complex olfactory stimuli as the mere tokening of their constituents. Since we can manipulate the concentration effect in configural mixtures to shift the overshadowing of the components, as well as change them into elemental mixtures, it suggests that these perceptual states are representationally complex. Veridical experiences of these mixtures present a complex entity that is not merely the summation of its parts. Moreover, the constituents are not identifiable as distinct tokens within the complex stimuli. Configural olfactory mixtures are not represented in a concatenative or functionally decomposable fashion. Rather, the best explanation of their representational structure only requires the most minimal notion of compositionality.

4.3.1.6 Olfactory imagery Olfactory imagery, whereby we can volitionally recreate the olfactory experience of a smell in the absence of sensory stimulation, further supports the claims that olfaction implements formative nonconceptual content. Originally, Algom and Cain (1991) showed that olfactory imagery mimics the veridical perception of odor mixtures. Our capacity for olfactory imagery (reviewed in Arshamian & Larson, 2014; Stevenson & Case 2005; Young, 2014), which mimics veridical perception, suggests that the compositional representational format is preserved from sensory and cortical processing through perceptual and cognitive states. A fuller discussion of olfactory imagery and its representational nature can be found in Young (2019c).

4.4 Explanatory Purchase of Formative Nonconceptual Content

Aside from providing precision in clarifying nonconceptual content as simply a matter of format relative to a system of representation, FNCC provides greater explanatory purchase by accounting for the puzzling discrepancy between our olfactory abilities (Young, 2019d). Humans are gifted at detecting and discriminating odors. Yet, we have difficulty identifying even the most prevalent everyday odors by name. One estimate of our olfactory ability to discriminate odors places the number around one trillion (Bushdid et al., 2014). While this number and the model used to derive it have been criticized (Meister, 2015), the latest estimate of the dimensional space for olfactory qualities using perceptible discriminations exceeds that of vision and surpasses our ability to name odors (Keller, 2017). When considering our olfactory performance for detecting and discriminating odorants, we are gifted. Yet, English speakers are notoriously bad at naming odors to the extent that if it were not so pervasive across the population, we might consider it pathological.[9]

We are remarkably bad at naming odors that we encounter on a daily basis. In experimental settings, humans can on average identify between 26 and 60 percent of the tested odorants by name, but this percentage varies across experiments and ages, and whether the odorant is presented to both nostrils. In one of the earliest studies on odor identification, Cain (1979) showed that of the eighty familiar odors presented, participants could only identify 60 percent.[10] In a further set of experiments Wijk and Cain (1994) showed that the percentage of odors correctly identified is related to age, with the highest rate of identification among young adults, where 53 percent correctly identified familiar odors. A follow-up study showed that odor identification varies with age—those between eight and fourteen years of age correctly identified 50 percent of odorants, which increased to 60 percent in the eighteen- to thirty-seven-year-old cohort, but slowly dropped to 40 percent in those between sixty-five and ninety years of age (Cain et al., 1995). However, these findings of 60 percent accuracy at odor identification are rather high. Other studies have reported far lower rates of identification, such as 26.5 percent (Huisman & Majid, 2018), 29 percent accuracy using one nostril, and 50 percent using unfamiliar odors (Jönsson et al., 2005). One reason Cain et al.'s (1995) reported identification rates might be so high is that they use familiar odors from everyday life, but when unfamiliar odors are employed, the accuracy of identification drops (Savic & Bergulund, 2004).[11] Additionally, Herz's (2000) findings that identification decreases with the use of only one nostril might explain the very low rate of identification in Jönsson et al. (2005). In a further set of experiments, Jönsson et al. showed that instances of failure of identification involve a failure of knowing more than a tip-of-the-tongue (or tip-of-the-nose in olfaction) phenomenon, which leads them to conclude that the lack of accuracy arises from a failure to identify the odor.

Failures of identification are not unique to olfaction. We have similar issues with identifying unfamiliar faces, which is attributed to the representational format of facial representation being holistic. In fact, it has been suggested that odor identification might be similar to facial identification in that they both use holistic representational formats (C. Murphy et al., 1991). However, in a rather ingenious set of experiments, Stevenson and Mahmut (2013) showed that odor identification is not completely analogous to facial identification. Even if we cannot identify a face, we still have access to associated semantic information about the face. Yet, if we fail to identify an odorant, we cannot access semantically associated information about the smell. Stevenson and Mahmut (2013) concluded that we have fully formed odor percepts in these situations. Yet, there is poor connectivity between olfactory perceptual centers and semantic memory processing within olfaction. Thus, our failure in identification is directly linked to an inability to access semantic information about the target smell.

Herein, we are presented with the puzzling discrepancy[12]: humans are gifted at discriminating odorants but pitiful at identifying odors by name. We can detect a vast quantity of odors (Bushdid et al., 2014),[13] and the dimensionality of our capacity for olfactory discrimination dwarfs that of vision (Keller, 2017). Yet, we accurately identify only a fraction of these (Cain, 1979; Engen & Ross, 1973). Identifying an odorant by name demands further cognitive resources than detection or discrimination because it requires generating the sematic label of the olfactory stimulus, suggesting that our poor performance is linked to semantic processing issues (Jönsson et al., 2005; Veramendi et al., 2013). Moreover, in the absence of correct identification, we cannot access associated semantic information about the odor (Stevenson & Mahmut, 2013), indicating that the discrepancy arises as an interface issue between olfactory processing and our conceptual semantic resources. We cannot simply explain our poor accuracy at identifying odors as deriving from being bad at smelling because quite the opposite is true. What will be argued is that our puzzling olfactory abilities do not merely derive from our linguistic abilities or connectivity issues. Comparing the format employed in olfactory processing as it projects to linguistic centers with what is known about the representational format of concepts generates a more comprehensive explanation of the puzzle than the current alternatives (for a discussion and criticism of the full range of alternative explanations, see Young, 2019d).

4.4.1 The Neural Realization of the Compositionality of Concepts

Our capacity to think thoughts with complex semantic content derives from our inferential sensitivity to a hierarchy of semantic content and rich interrelations between the associated constituents that generate the complexity we attribute to concepts. What is decisive throughout the literature on concepts is that either these entities have an internal structure that can be used to explain their complex semantic value, or we can refer to their constituents by way of their compositional formation and decomposition function. Regardless of one's theory of concepts, a necessary feature of concepthood is that we can potentially access the constituents that compose the complex representation when deploying the concept.

Only recently have we begun to identify the underlying cortical realization of the hierarchical structure required to implement these notions of conceptual compositionality. The cortical realization of concepts is accomplished through dynamic hierarchical binding across levels of neural networks, which generates coherent internally structured complex representations without an explicit concatenative syntax (Maye & Engel, 2012). Furthermore, the perception of visual objects depends upon hierarchically structured semantic resources. Visual object perception is refined using these high-level object features that are coded together with semantic representations (Clarke et al., 2013, 2015). These studies provide strong evidence that hierarchically structured semantic information plays an integral role in the formation of visual object representations with their paradigmatic compositional format. The aforementioned results might be relativized to vision, but there are similar experimental results for the cortical realization of the hierarchy of compositional structure in semantic linguistic processing.

Parts of the anterior temporal lobe (ATL) are key in processing conceptual representations (Binder, 2015; Clarke & Tyler, 2014; Clarke et al., 2015; Mion et al., 2010; Moss et al., 2005; Patterson et al., 2007; Tyler et al., 2004), which is in keeping with Olofsson et al.'s (Olofsson, Rogalski et al., 2013; Olofsson et al., 2014) research on the semantic processing areas of olfaction. However, the ATL's role in linguistic processing involves representing the syntactic organization and hierarchy of linguistic stimuli, while the angular gyrus (AG) encodes the semantic features and hierarchy that we would commonly refer to in explaining compositional thought. A series of studies have shown that the AG processes combinatorial semantics (Humphries et al., 2006), forms the neuroanatomical hub for conceptual combination (A. R. Price et al., 2015a, 2015b), and plays a causal role in the compositional integration of lexical semantic information (A. R. Price et al., 2016). However, all three studies indicate that the ATL does not process the combinatorial features of semantic processing required in implementing the semantic hierarchy that we attribute to the compositionality of concepts. Not only has it been shown that the cortical realization of conceptual sematic capacities requires a hierarchical structure of encoding allowing for compositionally complex semantic concepts that are sensitive to their individual constituents, but also that the ATL does not encode the semantic aspects of compositional representations for conceptual representations.

4.4.2 Explaining the Discrepancy Based on Cortical Connectivity

A recent line of research explains the discrepancy between semantic resources and olfactory processing as a connectivity issue. In a progressive series of experiments, Olofsson, Rogalski et al. (2013; Olofsson et al., 2014) and Olofsson and Gottfried (2015a) mapped the linguistic centers responsible for our ability to represent linguistically and report verbally our experience of odors. Identifying and naming a stimulus requires a three-part series of perceptually transducing the odor object, representing the stimulus, and generating a verbal report using the representation of the perceptual entity. Using this tripartite breakdown of the prerequisites of fulfilling a naming task, they began by first studying the verbal production of semantic tags for odors. Using a cohort of patients suffering from primary progressive aphasia, Olofsson, Rogalski et al. (2013) localized areas within the temporal pole (TP) and inferior frontal gyrus, which process the verbalization of odors. In a further set of studies, they showed that the TP in the ATL mediates the connection between olfactory representation in the piriform cortex and language centers (Olofsson et al., 2014). Based on the sequence of studies, they identify these areas as olfactory-specific centers for olfactory and semantic integration, which are dedicated to the linguistic mediation of olfactory representations.

Employing these studies, they note that there are only three relays between olfactory sensory processing and these cortical areas. The lack of connectivity leads Olofsson and Gottfried (2015a) to hypothesize that the lack of cortical relays yields a dearth of processing. Accordingly, they conclude that olfactory semantic integration is inferior because less processing has occurred to the sensory information by the time it reaches the ATL. Their theory tacitly assumes that the decreased cortical connectivity implies a decreased amount of processing, and this alone is sufficient to explain the puzzling discrepancy.[14] While their research is fundamental in unraveling our perplexing olfactory abilities, there is a significantly underexplored aspect of their proposal concerning the format of olfactory representations, which I will seek to complete below.

4.4.3 The Alternative: Representational Format

Our inability to identify smells by name is not only attributable to the lack of connections between olfactory processing and language centers. Rather, the location of these interconnections allows us to infer that the best explanation involves interface issues derived from an incompatibility of formats. The cortical connection that feeds into the olfactory language center identified by Olofsson, Rogalski et al. (2013; Olofsson et al., 2014) derives from the PC, whose encoding structure is distributed in a manner that is not reminiscent of the compositionality observed in the other perceptual systems.[15] Thus, it is not just the connectivity but also an incompatibility of representational formats that explain the puzzling discrepancy.

The format of stimuli transduction and encoding from sensory to cortical areas in the olfactory system as it projects to olfactory language centers does not require the explicit representation of each constituent within the complex representation, yielding a form of functional compositionality that does not readily allow the complex to be decomposed into its constituents, which facilitates efficient detection and discrimination but not identification. The format of olfactory stimuli transduction and encoding allows for an extremely efficient representational system that is sensitive to the vast range of variegated chemical stimuli (including complex mixtures) that we can detect and discriminate.

The reason we cannot identify odorants with ease is because our conceptual semantic resources operate using a hierarchical system of representation that is either concatenatively compositional or able to access the constituents using a decomposition function, while most of the time, the olfactory system employs a more minimal form of compositionality. The language centers of olfaction do not represent or parse the incoming projections from the PC in a manner that enables it to be readily employed by our conceptual semantic processing hubs. The modality-specific area of the cortex responsible for olfactory language convergence, the ATL, receives its input from the PC, which does not obey the strictures of classical compositionality or even decomposable functional compositionality. Moreover, the ATL has been associated with processing the syntactic aspects of compositional linguistic representation, but it is not implicated in processing compositional semantics. Future empirical research is required to support my thesis. However. explaining the puzzling discrepancy as an interface issue between incompatible compositional formats allows an integrated explanation across each stage of olfactory processing, as well as providing a unified explanation of why we are gifted at olfactory discrimination but bad at identifying smells.

4.5 Conclusion

We smell mereologically complex chemical objects that are unlike the ordinary visual objects we refer to on a daily basis, making it unsurprising that our experiential states about smells also implement an alternative format of representation. What has been shown throughout this chapter is that our olfactory states are representational and employ a nonconceptual format. The olfactory system implements a non-concatenative and non-decomposable form of compositionality that is preserved across sensory, perceptual, and cognitive states, such that olfactory experiences have robustly complex content deriving from their constituents. Yet, we cannot access these components within these functionally compositional states.

Using olfaction as our target domain, this chapter was able to provide a novel explanation of the nature of nonconceptual content dubbed FNCC. Not only is FNCC of theoretical interest, but it was also demonstrated that olfaction biologically implements such a format of representation. Additionally, this outlier format of representation provides a possible explanation to the discrepancy between our gifted olfactory ability for odor detection and discrimination and our almost pathological inability to name even everyday smells, given that our olfactory cortical processing hubs and our conceptual processing centers employ incompatible formats of compositionality.

What is of further interest is that despite smell and flavor perception being dissociable based on their difference in neural processing, olfactory qualities, and object of perception (chapter 3), nonetheless they both, to a large extent, neurally realized by the olfactory system, as is apparent based on their similarity in compositional formats. As noted in chapter 3, when we have an experience of flavor, the percept is as of a unified object that cannot be decomposed within synchronic experience into its independent sensory components, at least in the case of retronasal and gustatory qualities (somatosensory, thermal, and chemothesis properties are to some extent dissociable; Stevenson, 2009a). What unifies our taste of flavors is the functionally compositional yet non-decompositional format of the combinatorial representation of the sensory qualities of the multiphasic chemical entity placed within the mouth, including its odorous headspace. Over time, we may probe the flavorful object to access its components and/or divert attentional resources to examine its components using the full range of sensory systems within our headspace. Nevertheless, the object of perception is presented as a synthetically unified percept, thereby displaying the dominance of the olfactory system, even within our perception of flavor.

Despite olfaction's strengths and the use of examining the format of smell for a range of debates within philosophy, we must not lose sight of the fact that thinking about, conceptualizing, and naming smells is extremely difficult for most of us. If olfaction implements this odd nonconceptual format, then it is reasonable to wonder how our cognitive states are able to categorize smells if only in a quasi-conceptual fashion, such that we can represent a complex smell as being of the same kind across instances, slight changes in mereological composition, and contexts. Additionally, how do we use these cognitive categories to communicate meaningfully about smells throughout our daily lives? The next chapter turns to focus on how we think and communicate about smells.

5 Pondering Smells

We are awesome smellers! The sheer number of new and novel odors that we can detect and discriminate is simply outstanding in its scope. Current estimates are that we can detect and discriminate approximately one trillion different smells. Our sensory sensitivity allows us to detect infinitesimally small amounts of a single odorant against the background of a disproportionate background smellscape. And as surprising as it may seem, human olfactory perceptual acuity is on a par with if not exceeding those of other animals that we consider to be gifted smellers (e.g., bloodhounds). However, as noted in the last chapter, there is a disparity when it comes to the connection between our olfactory perceptual states and linguistic capacities that might be attributable to a difference of compositional formats. Nevertheless, we do think about smells, such that we can wonder if a smell we recognize from childhood, such as lilac, would smell the same on a different continent in a new and novel context. We successfully communicate and talk about our smell experiences. We ask relatives if the substitution of an herb or spice within a family recipe changes the complex odor profile of a dish. Despite our olfactory limits, we do think about smells as categorical entities and can communicate about them as such.

The general focus of this chapter is how we cognitively categorize smells. What is quite clear from the outset is that olfaction will not employ concepts in anything like the classical philosophical notion of propositions with necessary and sufficient conditions (Margolis & Laurence, 1999, chapter 1). Moreover, given the argument in the last chapter that the format of olfaction is conserved across levels of processing within olfaction, then it would be odd to talk about olfactory concepts. Nevertheless, it seems possible that perceptual processing can generate categorical cognitive representations ranging over instances of token experiences without concepts (Deroy, 2019). And although the prevalence and use of smell categories are meager within our daily lives, we nonetheless do think, imagine, and talk about smells as categorical kinds. Thus, the chapter explores how it is that we can characterize smells categorically, such that we can think about smells across multiple instances as being of the same kind and interpersonally communicate about smells. Put another way, what passes for an olfactory analogue of concepts in terms of cognitive smell categories that we use to think and communicate about smells?

Pondering odors encompasses more than just their olfactory quality. Our cognitive states about smells also encompass their intensity, valence, hedonics, and associations that might be mediated by cultural practices, mode of living, and functional usage within daily life. Thus, the first thing to consider is whether humans' poor ability to name odors is universal or an artifact of linguistic and cultural practice that are predominantly measured within English speakers. Section 5.1 reviews the literature that suggests the difficulty in naming odors is meditated by language and culture, which then sets up the question of what our linguistic practices can tell us about the representational format of smell categories. Section 5.2 picks up on this strand to show that the relation between our linguistic conceptual resources and olfactory representational states is symmetrical poor. We are bad at naming perceived odors, and verbal tags do not generally elicit olfactory representations of smells. The poor connection between language and olfaction provides further reason to think that if representing smells requires a complex system that is not a simple isomorphic mapping relation of odorant to odor, then how we cognize and categorize smells will also follow a nonlinear pattern. Section 5.3 employs the claim that olfactory cognitive states also use a holistic combinatorial format in providing a sketch of the representational format of smell categories. The section suggests that the most adequate method for studying smell categorization is experimental paradigms that make room for similarity ratings between smells as opposed to assuming a linear matching between odors and smell categories identified by a lexical tag. The section further explores what explains experts' abilities to form more nuanced smell categories that enhance their discriminative perceptual judgments. Despite the stinking relation between language and olfaction, section 5.4 then explores how language and expertise influence categorical ability. The chapter concludes that the format of smell is most likely preserved across sensory processing, perceptual, and cognitive states, which then sets us up to consider olfactory consciousness in the next chapter.

5.1 Naming Smell across Languages

Philosophers of the last century employed our linguistic abilities not only as a means to theorize about fundamental metaphysics but also as an access point to the structure of our contentful mental states. Thus, as noted in the last chapter, philosophers of mind attempted to ascertain the format of cognitive states by inference from the systematic and productive structure of language and our comprehension thereof. The last chapter built upon this strategy to argue that olfactory sensory, perceptual, and cognitive states employ a compositional format that is incompatible with that implemented by our linguistic systems and semantic accessible conceptual repertoire. Aside from providing an empirically tractable manner to assess the debate about nonconceptual content, formative nonconceptual content has the added bonus of providing a means of reconciling our gifted olfactory capacity for stimulus detection and discrimination with our poor ability to name smells. However, the claimed naming deficit has recently been called into question because it is predominantly relative to English speakers and Western cultures. Based on a growing body of research on smell (and taste) lexicons throughout the world, our inability to name odors might not be universal (Majid, 2021; Majid et al., 2018).

However, if the ability to talk about and name smells fluctuates across languages and cultures, what then explains the differences in olfactory prowess? What will be argued throughout this section is that the format of olfactory perceptual and cognitive states as being combinatorial and functionally compositional is still the most likely explanation of the format of olfactory categories across languages. Having established that there is a poor fit in the direction of perception to naming, the next section will then employ further research on odor simulation by smell vocabulary to show that the poor fit goes both ways. The symmetrically poor connection solidifies the driving claim of this chapter that cognitive smell categories preserve the functional compositional format of olfactory perception.

5.1.1 English Speakers Cannot Name Smells (Well)

WEIRDos (Western, Educated, Industrialized, Rich, and Democratic) are remarkably bad at naming odors, even those that we encounter on a daily basis. As a means of self-demonstration, Yeshurun and Sobel (2010) suggest performing the refrigerator smell test. Take any item out of the refrigerator and ask someone to close their eyes and name the item based on its smell. Most people cannot identify things as simple as peanut butter, ketchup, or mustard. Yet, the refrigerator smell test is an unfair method for assessing our olfactory abilities, since these stimuli are being presented at a low temperature that is not optimal for generating the volatility of the chemical compounds necessary for the perception of olfactory quality. By comparison, the refrigerator smell test is on a par with asking you to identify the colors of people's clothing in a dark movie theater. The do-it-at-home test is far from perfect. Yet, even under more ideal conditions sampling from the pantry cupboard, our ability to name odorants is poor.

While the evidence surveyed in section 4.4 paints a meager picture of our ability to identify smells, this is relative to the methods of asking participants to generate freely an exact name for the target odorant. Naming smells in this sense requires not only categorizing a smell as falling within a given class but also providing a label for the smell. The experimental measure used for naming smells is a particularly complex cognitive task, but it does not entail that we have either a deficit in describing smells or difficulty generating smell categories.

5.1.2 Difficulty in Identifying Smells Is Not Universal

The discrepancy between olfactory capacities for identification and discrimination is not universal. The difficulty in naming smells might be due to differences across linguistic communities and the cultural importance of smell. The ease of describing and talking about smell might be modulated by the placement of smell within a culture's hierarchical conception of the senses—the role of the prevalence of the use of smell language might modulate olfactory categorical abilities and their specificity (Majid et al., 2018). But our olfactory communicative abilities also have to do with a range of factors such as climate, the prevalence of smell terms in everyday life, the importance of cooking and cuisine to one's culture, the level of industrialization, and ecological lifestyle (Majid, 2021). What becomes apparent from research on how culture and language modulate our ability to think and talk about smells is that English and Western cultures seem to be outliers.

English does not have a designated odor lexicon (Majid, 2015; for contrastive views, see Olofsson et al., 2021; B. C. Smith, 2023), but cultures with a robust lexicon for smells that use these throughout everyday life display little to no discrepancy in identifying odors in comparison to colors (for a brief review, see O'Meara & Majid, 2016). Despite only showing 26.3 percent accuracy at odor naming, a recent study by Huisman and Majid (2018) suggests that the familiarity of the odor label also plays a modulating role in increasing naming accuracy. Thus, they conclude that odor-label frequency also needs be taken into account to explain the difficulty in naming odors. For instance, Maniq has a rich culture of talking about smells using a two-dimensional odor lexicon that does not extensionally refer to natural kinds but picks out common odor qualities. Additionally, the Aslian community of Jahai has a dedicated lexicon of smell terms and categories that are integral to their daily lives. Moreover, when compared to English speakers, they show no deficit between their descriptive capacity for colors and odors (Majid & Burenhult, 2014) to the extent that it has been argued they do not show a deficit in naming smells. In a more recent study, Majid and Kruspe (2018) go on to attribute the inability for naming smells as a culturally contingent fact that is related to subsistence mode, such that it is not seen in their subset of Aslian communities who are hunter-gathers, but it is seen within those whose subsistence depends more on agriculture and trade. Thus, the discrepancy is not a universal phenomenon but rather partially dependent on language, culture, and mode of subsistence.

Even more recent research has shown that aside from small nonindustrialized languages that show a robust capacity for describing smells, it turns out that smell lexicons can be quite vast, depending on the target language being studied. Thai, spoken by tens of millions of people, has a robust set of vocabulary for smells that is best characterized not in terms of identifying smells with source objects but rather as abstract categories ranging over classes of exemplars (Wnuk et al., 2020). Additionally, the fineness of grain of the terminology used to characterize smells has been shown to shift between generations based on changes in lifestyle and Westernization. For example, de Sousa (2011) documents the changes in Cantonese across generations attributed to modernization in Hong Kong and Macau, such that verbal terminology became increasingly fine-grained for making visual distinctions at the cost of loss of grain within olfactory categorizations.

Based on this edifying research on olfactory lexicons across cultures, it is safe to conclude that olfactory language is not universally used except on rare occasions, that it allows humans to categorize smells with a specificity that tracks daily usage and the importance of smell and cuisine, that it can be a robust portion of a culture's lexicon, and that it can even be used to generate olfactory specific metaphors (O'Meara & Majid, 2020). Despite showing that non-Westernized languages can communicate, categorize, and identify smells well beyond their English-speaking counterparts, one may wonder what facilitates this difference and what can be learned about the representational structure beyond their status as classes ranging over exemplars, as is the case in Thai's smell categories.

5.1.3 Stinking Methods

As a pedantic stinking philosophical point, it should be noted that there is a methodological inconsistency between the measure employed by Majid and Burenhult (2014) and Majid and Kruspe (2018) and studies of odor naming in English. Odor naming as reviewed in section 4.4 requires participants to generate a lexical tag for the presented stimulus freely, whereby accuracy is determined based on a matching relation between the exact name of the stimulus and the subject's report. The task is cognitively difficult, as it requires the subject to attend to their perceptual experience, recognize the target smell, and then freely recall the name. To understand the difficulty in this type of smell identification task, it should be noted that contrastively when using Sniffin' Sticks in clinical settings to establish a patient's odor identification ability, these come with three possible options for the patient to select from. When the difficulty of the task is attended to, it is still surprising that English speakers have such difficulty with the task, given that similar visual experiments often result in a ceiling effect.

The reason for noting the disanalogy between the odor identification tasks used for studying naming in English and cross-cultural studies is that it oversells the claimed results as being about naming when in fact it might be fairer to say that they show divergence in identification as measured by descriptive analysis. The method employed in both Majid and Burenhult (2014) and Majid and Kruspe (2018) compares the description of the participant to the experimental cohort in terms of (a) descriptive agreement, (b) length of utterance, and (c) type of response. The descriptive analysis is not the same method employed in the studies of English speakers, which requires the participant to generate an accurate name for an odor freely. However, that is not to question the validity of the cross-cultural comparison of Majid and Burenhult, as their measure compared English to Jahai speakers on points (a)–(c) for descriptions of colors and odors under the assumption that smells are qualities of objects and not objective perceptual entities (an assumption criticized by Olofsson & Wilson, 2018). While there might be less of a discrepancy in the descriptive abilities of non-English speakers between their capacity to describe colors versus smell accurately, it turns out that even English speakers are decent at describing smells with their limited smell lexicon.

Descriptive methods for identifying smells, in fact, show improvement, as might be noted in the non-forced choice paradigm in English as well. People can improve their capacity to name odors in a short period of time through training. Cain (1979) claims to have shown an increase from 60 to 77 percent accuracy over the course of four sessions spaced two days apart. In the retesting session, the task reused the linguistic labels from the previous session, thereby generating a measure of matching and recognition across trials and not the self-generated naming task that was used in the initial conditions. Similarly, Sulmont-Rosse (2005) showed that if individuals reuse their self-generated linguistic labels across trials, their odorant recognition increases, but this does not demonstrate an increase in odor naming. Directly contradicting Cain's (1979) finding, Jehl et al. (1997) showed that verbal encoding enhances the encoding of odorants for short- and long-term memory in a manner that increases detection and recognition. Yet, identification decreased across trials. In short, the literature paints a varied picture of our ability to identify smells that nicely brings the troubled connection between olfactory perception and our ability to identify smells in terms of either descriptions or names. However, what is left to ascertain is how the relation works in the opposite reaction. Is the troubled relation symmetrical, such that generating names for smells is difficult and eliciting olfactory cognitive categories using linguistic representations is equally hard?

5.2 Olfactory Categorization beyond Perception

Smell perception requires a highly dynamic system of odorant encoding that can accommodate the vast complexity of the stimulus across at least the parameters of odor quality, pleasantness, and intensity (Keller &

Vosshall, 2016). Most of the things we consider as smells range over multicomponent odorant mixtures with mereological structures, such that the dimensionality of our mental quality space of smells is many orders beyond the dimensionality of the sensory qualities of other sensory modalities (Keller, 2017; Young et al., 2014). With the representational complexity involved in just transducing what we perceive as smells, it seems fair to predict that our cognitive categories employed to think and talk about smells would have at least the same level of complexity and preserve the format employed within olfactory sensory and perceptual states. The following section aims at strengthening the parsimonious inference that olfactory categories employ a representational format that is combinatorial and at least similar (if not the same) in compositionality as the other olfactory states documented in chapter 4.

From the outset, it should be noted that this section presupposes the general working assumption that is supported by evidence surveyed in chapters 2 and 4 that the olfactory system should not be considered as a linear sensory system with strict isomorphic mapping relations between odorants and their corresponding percepts, including sensory qualities such as olfactory quality. Additionally, it is dubious that we should expect smell to obey strictures of classical conception of concepts (Margolis & Laurence, 1999, chapter 1). From what follows, it will become clear that our smell categories form a reliable means of mapping the odorants within smellscapes to olfactory cognitive states that range over kinds of smells, including their valence and emotional salience, as well as other associative information.

5.2.1 Olfactory Categories Are Best Probed by Comparison

The limitation in identifying smells by names (although perhaps not universal) suggests that individuating smells into distinct categorical kinds might require employing a different behavioral experimental method. In general, if smells are mereologically complex perceptual entities whose representation requires a holistic compositional format, then our cognitive states concerning these underlying percepts must also be probed using a similar method. To this end, sorting paradigms are often employed by sensory scientists looking to uncover how humans group smells and flavorful entities. The method allows for nonlinguistic measurements of how we generate categories ranging across disparate yet similar consumer products, such as food and beverages. Sorting paradigms move beyond linguistically mediated decisions about shared similarity or identity based on lexical tags by asking participants to sort samples into groups based on perceived similarity. Often, in this guise, participants must probe the mereological components of each item for recognizable features that can then be sequentially compared to the compositional structure of other complex smells.[1] Understanding our cognitive categorization of smell as preserving FNCC is supported by research on sensory sorting tasks of wine, whiskey, and coffee, and can be exemplified by a study conducted by B. C. Smith et al. (2017) that showed blended and single-malt whiskeys are not discriminable as separate perceptual categories by either experts or novices. Rather, whiskeys are grouped in terms of similarities of components within configural representation.

5.2.2 Categorizing without (Local) Concepts

Depending upon the target theory of concepts, it is arguably the case that olfactory cognitive categories might be conceptual. However, even if philosophical orthodoxy with classical concepts bordering on propositional essences is your thing, it might nonetheless be possible to consider olfactory perceptual and cognitive states as generating categories, even in the absence of conceptual status. Deroy (2019) makes a powerful argument based on the processing speeds and lack of conscious mediation of olfactory percept creation that we should allow for the theoretical possibility of odor categorization without concepts. But even if smell categorization is possible without concepts, we still need to explain how we employ these to think about and describe the varieties of kinds of smells, what individuates these categorical kinds, and how their compositional structures permit comparisons of similarities within and across types of smells.

To the best of my knowledge, the most recent and comprehensive treatment of the nature of olfactory conceptual spaces is provided by Jraissati and Deroy (2021). Their careful analysis of olfactory quality space yields the implication that, given the non-closed nature of the perceptual dimensions of smells, as well as the variability in classification across individuals, cultures, and linguistic communities, smell cannot generate a universal and global conceptual space with the same level of fine-grained distinctions possible as the conceptual space of colors (or visual objects more generally). Despite this disparity with vision, they conclude that smell can generate cognitive categories within a course-grained holistic conceptual space.

And based on linguistic mediation, they further allow for local fine-grained concepts within expert communities relative to their sensory domain of expertise. I generally agree with their arguments and evidence that our smell categories are generally synergistic and expert training can produce localist categories, but I think the formative structure of these is conserved from our olfactory sensory and perceptual states. What will be argued is that all olfactory categorization occurs within a compositional format that is combinatorial coarse-grained, such that the best explanation of expertise is not lexical prowess but rather the training that facilitates the acquisition of greater sensitivity to the salience of mereological patterns within the complex composition of smells.

5.2.3 Olfactory Simulation (or Why Language Stinks at Eliciting Smells)

The relation between our perceptual capacities and linguistic abilities might be fraught not only going from perception to linguistic access and reports but also in the opposite direction as well. Not only are we bad at naming what we olfactorily perceive, but verbal cues and linguistic names serve as a poor means of eliciting olfactory percepts as determined in terms of their cortical realization. Evidence for our capacity for olfactory imagery—that is, eliciting a sensory percept of a smell volitionally in the absence of any odorants—is meager in comparison to other modalities, but it is quite possible. Moreover, it is arguably the case that olfaction implements the same compositional format during imagery experiences as veridical perception, such that olfactory memories re-eliciting smell experiences are not interfered with to the same extent as visual or auditory autobiographical memories (Young, 2019c). The growing body of research on volitional smell imagery suggests this is due to the activation of perceptual systems and not mediated linguistically, which is corroborated by the study conducted by Speed and Majid (2018) displaying a lack of evidence for olfactory simulation.

Mental simulation is the phenomenon whereby stimuli such as linguistic terminology either names or descriptions elicits or activates (i.e., simulates) perceptual representations. An earlier study by Gonzalez et al. (2006) using fMRI readings found that verbal tags for odors elicited activations of odor representation simulation in the primary olfactory areas (piriform cortex and amygdala) but not the secondary olfactory cortical processing hubs (orbitofrontal cortex). However, the timescale (three seconds) they employed for stimulus presentation cannot rule out the possibility that their results might be attributed to mental imagery instead of simulation. Moreover, results from a more recent study conducted by Speed and Majid (2018) indicated that in comparison to vision and audition, olfaction seems to be the exception to embodied theories of language, with their central tenet that language elicits a mental simulation of the underlying perceptual representational states. Not only did their findings not indicate the ability for linguistic representations to elicit smell percepts, but they also showed that olfactory representations are best conceived of not as fine-grained categories ranging over the particularities of specific smells. Their results indicate that odors are mentally represented in a coarse-grained fashion because both fine-grained tags and mere similar lexical tags were equal in their word-recall memory tasks. Nevertheless, they do note a tighter and more fine-grained relation between linguistic representations and smells when it comes to higher-level smell properties such as ratings of intensity and pleasantness, as these might be mediated by associative learning and connections to nonolfactory cortical processing. Based on their finding that olfactory states employ a coarse-grained format for the representation of smells (i.e., smell categories individuated in terms of olfactory quality), Speed and Majid concluded that "olfactory language is not grounded in primary perceptual representations even at a coarse grain" (p. 377).

A number of conclusions can be derived from Speed and Majid's (2018, 2020) research about the structure of olfactory categories. First, olfactory categories are neither primarily constructed nor influenced by language. Second, these cognitive smell categories are not fine-grained, which allows them to range ambiguously over multiple similar sorts of odorants with similar olfactory qualities. Lastly, when smells are considered as a complex combination of odor quality and valence, these show differences in terms of both the degree of simulation and representation grain that can be attributed to the hedonics being associative and influenced by nonolfactory systems. Despite their findings neither fully elucidating what representation grain is nor how it is precisely measured,[2] the results are in keeping with the claim that olfactory categories are not fine-grained, lending support to Jraissati and Deroy's argument, as well as the conclusion of the last chapter that the olfactory system generally implements a compositional system of representation whose constituents need not be explicitly represented, thereby allowing that one complex representation might holistically range over an array of similar yet nonidentical mereological complex smells in a way that explains our smell categories' course-grained and ambiguous nature.

Smells might not be represented in a similar fashion to colors and perhaps other visual objects, but they do employ a similar holistic representational medium to that of face perception, as noted in chapter 4. The similarity in perceptual format between smells and faces can also be located in the representational format of their corresponding memories. Kärnekull et al. (2015) studied the similarities between forgetting smells and faces across time. Their study indicates that our long-term memory for odors follows a similar pattern of forgetting as that of faces. Yet, we are still better at remembering faces over long periods of time. What is interesting about this study for my purposes is that it shows that both target domains employ a similar holistic representational format that is combinatorial, thus suggesting that we represent both faces and smells as complex mereological patterns of the component parts not in a summative fashion but rather in a functional compositional format. What is left to ascertain is the role language plays in the odor categories of experts.

5.3 Olfactory Acuity and Categorization Is Primarily Driven by Perceptual Experiences

Olfactory perceptual acuity and, similarly, olfactory expertise are mostly mediated by experience-dependent representations rather than categorical representations mediated by our linguistic conceptual repertoire. Olfactory perception is primarily enhanced through an increase in the number of stimulus presentations and experiences. Furthermore, odor categorization is predictably similar across cultures in accordance with the prevalence of an odorant within the environment together with the judged similarity of the perceptual qualities.

The development of perceptual abilities in modalities such as vision and audition lags behind that of olfaction. The olfactory system is fully developed and functional in utero and is responsible for an infant's ability to identify its mother (R. H. Porter & Winberg, 1999; Russell, 1976), as well as distinguish relatives from strangers (Schaal et al., 2020). Children's olfactory capacities are fully developed by the age of three and are comparable to that of an adult in terms of odorant threshold sensitivity and hedonic judgments. While there is some difference in odor threshold sensitivities, this is most likely attributed to the plasticity of the olfactory system's adaptation to ecologically important stimuli. Furthermore, while some studies have shown that children's ability for odor recognition and identification is inferior to that of an adult, when linguistic competence and overall vocabulary are controlled for, this apparent difference disappears (Lehrner et al., 1999).

The process of maturation is unlikely to be a major influence on olfactory perceptual acuity. Perceptual acuity in the first few years of life might depend upon the development of the olfactory system, but a more plausible explanation, in keeping with the other evidence that follows, might be in accordance with the number of olfactory experiences required for the creation of a robust mental quality space on a par with adults. Maturation does not enhance acuity, but deterioration plays a role in the loss of perception with aging, starting at about the age of forty (Dulay et al., 2008). At around the age of sixty-five, olfactory acuity begins to decrease markedly, and by the age of eighty, there is a noticeable olfactory deficit in 75 percent of the population (Frank et al., 2003; for a recent review, see Olofsson et al., 2021).

Perception is doubtlessly influenced by our conceptual abilities and our linguistic practices utilizing vocabulary to name perceptible properties. Olfactory perceptual acuity is influenced not only by verbal mediation in terms of learned linguistic tags, but also, to a large extent, by the number of exposures to an odorant. In a classic set of experiments, Rabin (1988) demonstrated that increased exposure to an odorant improved the subject's ability to discriminate the target from others. Moreover, in a second experiment, Rabin demonstrated that the familiarity of an odorant allowed an enhanced discriminative capacity for similar odorants.

Since perceptual acuity increases even for identity and naming in accordance with familiarity of the odor (Homewood & Stevenson, 2001) and practice (Cain, 1979), it is worth considering how olfactory acuity is mediated by memory. Our almost pathological inability to name odors (Olofsson et al., 2013) has led many to question the format of odor memory. The underlying mechanism and process are still being investigated, but growing evidence suggests that olfactory memory is not mediated by linguistic tags or verbal coding. Odor memory is possible without verbal mediation (Møller et al., 2004). Olfactory coding and experiences are neither linguistically formatted nor dependent on language processing in the same way as vision (Goodglass et al., 1968; Herz, 2000).

Our perceptual ability for olfactory discrimination increases with training and exposure. Odors that are more familiar are easier to discriminate (Jehl et al., 1997). For example, perfume shopworkers have an increased ability to discriminate odors. Yet, their stimulus detection threshold and ability to identify odors from a list of descriptors is not enhanced (Hummel et al., 2004). These results suggest that peripheral sensory plasticity or an increase in linguistic tags are not the determining factor in their increased discriminative ability but rather some manner of perceptual sensory templates. Olfactory memory enables our capacity for perceptual discrimination in a manner that is not linguistically driven but improves through increased quantity of conscious or unconscious experiences that the subject undergoes. There is some indication that enhanced perceptual acuity in olfaction can even occur independent of conscious mediation and enhance our discriminative abilities. Increased presentation of an odorant, even subliminally, can generate further olfactory abilities for detecting and discriminating that specific stimulus. Wysocki et al. (1989) demonstrated that merely increasing the presentation of a stimulus enables a subject to gain the ability to detect and discriminate an odorant they were previously unable to smell.

Perceptual acuity improves with an increase in olfactory experiences and training that is best not attributed to linguistic mediation, which can be corroborated by research on wine experts. Wine experts outperform novices at odor discrimination (Bende & Nordin, 1997; Melcher & Schooler, 1996; Solomon, 1990), and their increased ability results from greater perceptual skill rather than verbal or descriptive resources (Parr et al., 2002). Parr et al. (2002) showed that the experts had an enhanced ability to recognize odors but did not outperform novices in terms of either their sensitivity threshold for odorant detection or the verbal memory task. When this result is combined with their findings that odor naming and odor recognition were not positively correlated, it provides reason to think that increased olfactory recognition and discriminative acuity do not depend upon enhanced linguistic tags or semantic descriptors.

Further results indicate that wine experts can more accurately discriminate between two varieties of wines, as indicated by their ability to sort samples correctly into respective groups (Ballester et al., 2008). Moreover, the results showed intersubject convergence of the experts on their judged typicality of each variety of wine, which was interpreted as indicating the experts delineated each kind of wine in terms of its perceptual characteristics. However, Ballester et al.'s (2008) results are neutral regarding whether the increased perceptual ability and judgments of typicality should be attributed to an enhanced perceptual strategy that is more analytic and focuses upon the perceptual qualities of the stimulus or an enhanced descriptive repertoire that would allow greater discriminative ability. A later study, which succeeded in training wine experts to detect and discriminate between key sensory characteristics using simple sensory training (Tempere et al., 2012), suggests it is the former. By merely exposing experts to key odorous wine compounds, Tempere et al. (2012) increased the experts' perceptual discriminative abilities. They were able to lower experts' detection thresholds through increased exposure to the key compounds in a fashion that allowed further discrimination within a group of qualitatively similar perceptible properties.

The enhanced discriminative abilities of perfume experts might be partially influenced by increased descriptive resources or linguistic labels, but the greater determinate is the actual number of experiences (Gilbert et al., 1998). In an experiment using a sorting paradigm, Veramendi et al. (2013) showed that perfume experts and novices mostly overlap in their odor categorization as determined by their sorting of perfumes in terms of perceived similarities and differences. And while the perfume experts were more parsimonious in the number of groupings, this was not statistically significant. Despite the experts' more exacting usage of linguistic descriptors for their odor groupings, the perfume experts' categories were mostly similar to those of the novice group. Not only did the perfumers' enhanced semantic repertoire and linguistic tags show no marked difference in their odor categorization, but where there were some differences in terms of their judgments of similarities and difference, that was best explained by the sheer number of experiences of the experts (Veramendi et al., 2013). Thus, even this slightly increased parsimony in odor categorization is best attributed to the number of experiences rather than linguistically mediated conceptual sophistication. These results indicate that olfactory acuity improves with the number of olfactory experiences in a manner that does not depend upon the maturation of the olfactory system or the nature of linguistic representation. Humans' discriminative abilities increase in accordance with the overall perceptual quality space of olfaction that is arguably not mediated by linguistic or verbal coding.

A key test for the claim that olfactory perceptual acuity is mediated by experience and not linguistic conceptualization or verbal coding would be cross-cultural comparisons of odor perception and categorization. Linguistic conventions and conceptual naming strategies differ between cultures. Yet, there is great deal of overlap in overall odorant categorization as determined by odorant sorting experiments using judged perceptual similarities (Chrea et al., 2005a, 2005b). Different cultures sorted the olfactory samples into similar groupings that were not consistent with their groupings of the odor labels that would be associated with the olfactory samples presented in the odorant sorting task. The odorants and their associated names not being categorized in the same fashion arguably demonstrates that the verbal labels did not determine odorant grouping. Additionally, the differences between cultures in their placement of odorants within groups displayed a familiarity effect. Cultures that were more familiar with an odor categorized it similarly, thereby showing that the number of exposures was the greatest indicator of olfactory discriminative ability for odor categorization in the odorant sorting task.

5.4 Categorization Can Be Enhanced Relative to Limited Target Domains of Expertise

Our olfactory capacity to categorize smells is primarily driven by increased exposure to smells, the importance of smells within our daily lives, and the amount of time spent conversing about odors, as well as the robustness of our lexicon devoted to olfactory sensations. Language helps us to acquire and form odor categories. But it is not the primary driver of our capacity for categorizing smells—language only enhances olfactory abilities relative to ecological need and expertise relative to its narrow area of abstraction. To show this, this section examines what explains the enhanced abilities of olfactory experts.

Language undoubtedly plays a role in odor categorization, but what will be argued is that it is primarily in acquiring a rubric for learning how to group odorants as having similar features. Recently, Vanek et al. (2020) showed that the use of novel verbal labels paired with odors helps form odor categories, such that merely using language tags that are not related to the sensory characteristics of the smell helps facilitate learning odor categories. However, their methods do not allow us to claim with certitude that these are robust odor categories, as they themselves admit the alternative explanation of the results as just learned superficial pairings cannot be ruled out. Yet, there are a range of studies that show verbal and written labels improve our ability to smell. Lyman and McDaniels (1990) showed that odorant encoding is improved when both verbal/linguistic and visual routes are also used to encode an odorant. Employing two extra encoding methods increases retrieval rates, making it likely that odorant encoding converts more resources from vision and semantic centers, thus increasing odor encoding and memory. Improvements in smell recognition and identification might be attributed to converting resources from visual cortical processing, which might sound implausible based on the speed of neural plasticity this would require. However, Qu et al. (2016) showed neural plasticity within two days of learning de novo odor–visual categories that required the use of both olfactory and visual cues. All indications from the experimental literature are that linguistic prowess does help for discrimination, detection, and recognition, but from the outset, it has been noted that we are gifted in this regard. While it is plausible that increasing our odor lexicon and verbal descriptions of smells might increase olfactory identification, even highly trained wine and coffee experts whose livelihood depends upon their use of specialized odor lexicons do not show increased accuracy in olfactory identification (Croijmans & Majid, 2016).

Occupational sensory experts can generate stable categories that can be used to identify similar and repeatable types that range across perceptible instances. Given their professional needs, it is unsurprising that their expertise presents itself in terms of their consistency in categorizing odors and flavors as of a particular type across multiple presentations. Experts are more accurate in sorting sensory items within their domain consistently into shared categories across presentation, as well as being more consistent in their interindividual groupings of sensory items into categories. Additionally, experts' descriptions of sensory qualities are more structured and concise and show greater consistency across experts. Lastly, their descriptions show greater precision in descriptive length (for a fuller review, see Majid, 2021).

Despite these enhancements in their holistic categories relative to their domain of expertise and descriptors thereof, a survey of the literature on olfactory expertise shows that experts do not have an increased ability for olfactory categorization in general, their increased lexical repertoire only mediates increased recognition abilities within their domain of training, and they do not have enhanced detection and discrimination or general olfactory perceptual acuity (reviewed in Majid, 2021). Building upon their research on wine expertise, Croijmans et al.'s (2020) study of wine experts showed that they are better at recognizing wines from memory, but this does not transfer to wine-related or common odors and is not verbally mediated (Croijmans et al., 2021). They note that having an exacting lexicon relative to a narrow range of sensory qualities might explain how linguistic mediation is required to initially acquired expertise, even if it is not a causal factor in recognition memory for odors. Not only is this evidence in keeping with Jraissati and Deroy's localist conception of olfactory concepts, it goes some ways toward explaining why experts show similarity of linguistic descriptions if trained using an exacting lexical system of descriptors used to categorize aspect of smells within a narrow range of odors.

So, what makes experts better within their area? Experts do not have enhanced linguistic-mediated smell concepts or olfactory abilities (detection, discrimination, or identification). Rather, I would speculatively suggest that their superiority within a limited range of odors is generated from greater recognitional acuity in teasing apart configural relations between parts of the sensory individuals within their domain of expertise.

Experts are confined to the same type of functional compositional format of olfactory representations but are better at recognizing patterns and salient components. Employing the same holistic format and sensory space of olfactory qualities that compose mereologically complex smells, experts are better at recognizing, for example, wines based on their greater familiarity and training to recognize the combinatorial complexity of the target stimulus within this domain (James, 2018). Given the configural representation of wines, their training allows for greater recognition based on holistic pattern recognition, which is borne out by the need to probe the wine perceptually across multiple sampling behaviors. The compositional format and our limit in perceiving the parts of a smell make it essential for experts to tease apart the component parts across multiple samplings to gain insight into the identity of the wine based on its compositional complex structure. The format of smell generates synergistic percepts with complex holistic representations, such that both experts and novices are on par relative to the sensory and perceptual abilities. What sets the expert apart is their training with a lexical framework that allows them to latch onto important features of the complex odor and then, through practice, better recognize the holistic representation of distinct categories. Additionally, their training improves their capacity to sample the sensory properties of the smell serially in search of these recognizable features.

Olfactory experts' perceptual acuity is not enhanced in terms of detection, discrimination, recognition, or identification of smells in general nor are their cognitive categories generally more nuanced or robust. Experts are not better at perceiving smells, but they have an enhanced ability relative to recognizing odors within their range of expertise as falling within more parsimonious smell categories that have great coherence intra- and intersubjectively. Moreover, their linguistic capacity to describe smells only relative to their domain of training displays that they most likely do have a unique set of specialized olfactory categories that can be deployed based on their training, which is in keeping with Jraissati and Deroy's model of localist olfactory categories. However, my account and the evidence surveyed above runs contrary to Barwich's (2020, chapter 9) claims regarding olfactory experts' differences in terms of the perceptual structure of expert noses and their finer nuances in perception. Our accounts of olfactory expertise differ on both matters. To be exact, according to Barwich, experts have an increased fineness of conceptual grain, whereby their categories are themselves more detailed in content, thereby enabling greater perceptual acuity. Contrastively, I think that, based on the evidence and arguments above, a better explanation of experts' abilities and cognitive categories is in keeping with the general representational structure of olfactory perception and cognition as preserving the functional–compositional format. Thus, olfactory experts' cognitive categories are specialized and localized to their narrow target domain, such that they might have a unique set of categories, but the representational grain is no finer than any other conceptual categories that we deploy in thinking and communication about smells. Finally, their perceptual capacities for smelling are not enhanced. They are documented as having an improved capacity for odor recognition that is arguably cognitively mediated via these narrow local olfactory categories rather than a perceptual enhancement.

If you will indulge me in an odd analogy that highlights my claim regarding experts' enhanced recognitional abilities, smells are like barcodes whose mereological complexity does not stand out to us as composing distinctive categories, despite our ability to see them as different, but through training to recognize certain groups and daily practice staring at barcodes, we could develop an expertise seeing their unique identities, such that their course-grained identities would become more salient to us in terms of the pop-out effect of the configuration of their components. Before concluding, it is worth calling attention to the fact that the scope of expertise under consideration ranges over both smells and flavors, which is fitting, given my earlier arguments that they both employ a functional compositional system of representation. Moreover, these findings regarding smell expertise corroborate that similar categorical mechanism are at play in both smell and flavor perception, with the latter being enhanced based on its conscription of additional cortical encoding and processing centers.

5.5 Conclusion

The representational format of smell is preserved across sensory, perceptual, and cognitive states, such that our smell categories are best conceived of as holistically structured representations that ambiguously range over coarse-grained similarities spaces including the olfactory quality of a smell, as well as its valence, hedonics, intensity, and associative states. Our smell categories allow for ambiguity in classing similar smells based on their patterns of resemblance between their mereological complex functionally compositional structures. The connection between our linguistically mediated conceptual systems and smell is symmetrically poor, as demonstrated by our deficit in odor naming (not descriptive identification), as well as the lack of evidence for smell simulation using linguistic labels. Thus, smell categorization is not predominantly mediated by our linguistic conceptual repertoire.

The sheer range of odorants, olfactory mixtures, and sensory qualities that the olfactory sensory system needs to track entails that it requires a dynamic means of representing the full range of perceptible objects. To counter this dynamic complexity, the FNCC representational format seems in an odd way parsimoniously fitting. It allows us to detect, discriminate, and recognize a vast number of smells that exceed what can be accommodated by our limited visuocentrically mediated linguistic conceptual resources that cannot grasp the nearly unbounded nature of smell. However, the format comes with its own limitations. We cannot elicit smells easily using linguistic descriptions and have great difficulty using linguistic mediums to name what we are smelling. Thus, it has been argued based on the empirical evidence reviewed that our cognitive categorization of smells is ambiguous across similar smells, making the best means of probing our conceptions of smells as requiring holistic dynamic methods that do not assume a linear matching relation between representation content and lexical tags.

While the vast majority of smell research occurs across Western cultures and predominantly English speakers, the arguments above show sensitivity to cross-cultural research, such that it seems safe to conclude that even for cultures with robust odor lexicons, their smell categories do not simply pick out ordinary source objects but range over exemplar classes of sensory olfactory qualities (e.g., Thai). Perhaps the course-grained structure of olfactory categories is not universal, but it was further argued that even experts might not have a more fine-grained conceptual space for the odors of their target domain. The difference between experts and novices is not grain of categories but rather recognition ability for patterns. Their training lexicon, which is focused on a narrow range of perceptual objects, allows for greater attention to key perceptual features, which facilitates acquisition of greater pattern recognition. Their linguistically mediated training allows them to form an initial set of categories, together with means for practicing different perceptual modes of probing the stimulus to tease out its complex sensory structure and combinatorial features. Both probing for reliable component structures of a complex and perceptual practice generate increased accuracy in recognizing the holistic representations as categories. However, experts are still stuck probing the stimulus multiple times to access recognizable patterns within the smell. We can ponder smells in a manner that is rather unique to the olfactory system as facilitated more by perception than our linguistically mediated conceptual repertoire.

More experimental work is needed to explore the structure of olfactory cognitive states further in order to substantiate my speculative conclusion fully, but while we await future research on smell categories, let us turn our attention to the relationship between smell perception and consciousness, which will be the focus of the next two chapters. And to prime the reader, there is a prima facie issue lurking about how to study our consciousness of smell if the amount of olfactory information processed by smell vastly exceeds our linguistic capacities, which seem so essential for introspective subjective reports that are often take as the gold standard in consciousness studies.

6 Unconsciously Smelling

Smells are ubiquitous. They envelop us and overlay the qualitative character of our daily lives. Although we may neglect their influence on our behavior and fail to take them into account in theorizing about concepts, consciousness, and the formative structure of our mental states, nevertheless they generate such a profound qualitative character that when our access to them is lost, our lives become duller, less enjoyable, and often downright depressing. We might not be continually consciously aware of the smells traversing our nostrils, but it will be argued that all olfactory sensory, perceptual, and (perhaps) cognitive states have a qualitative character to them, even when we are subjectively unaware of them as such. What will be shown is that olfactory consciousness provides a novel means of demarcating different kinds of consciousness. By adapting the distinction between access and phenomenal consciousness in an empirically tractable fashion, a contrastive understanding of the relationship between conscious awareness and phenomenal consciousness will be provided.

All of the different aspects of smell that we have noted as setting it apart from many of the dominant views across debates within philosophy will also be its source of departure in generating an understanding of olfactory consciousness. The prima facie issues in generating an account of consciousness in general that adequately captures smell are multifarious: the olfactory system has a divergent anatomy; the dimensionality of olfactory qualities vastly exceeds our default visual sense's dimensionality; we generally are not attentive to smells; introspective access as a method for theorizing about smell is inadequate; and olfactory states employ a format of representation that is not easily mediated or accessed by our conceptual repertoire as mediated by the linguistic system. However, our experience of smells pervades our waking consciousness, acts as a foundational component of the quality of our lives, and modulates some of our most profound behaviors. In short, it will be argued that we need to modulate our thinking about the relationship between awareness and phenomenal consciousness to encompass smell because our olfactory states have a qualitative character that influences our behavior, even when we are not subjectively aware of smell.

6.1 Definitional Matters

Debates regarding the nature of consciousness and its taxonomic kinds often become conceptually murky. So, it is best initially to clarify my use of terminology before entering into a discussion of the empirical evidence in support of each kind of consciousness. Pre-theoretically, being aware signifies that we can subjectively report undergoing the experience, being in the relevant state, and the content of the state. For my purposes, I shall use "awareness" to designate states in which the subject can report being in state *S* with content *p* (or if you prefer, they are conscious of undergoing experience *E* that is of, or about, object *x*). Awareness, I shall stipulate, can be understood separately from qualitative consciousness, such that an organism is in a qualitatively conscious state when there is something that it is like for it to undergo experience *E*, which is distinguishable from undergoing experience *E**, and moreover the subject need not be aware of being in state *S* (i.e., undergoing *E*) or of state *S*'s content *p* (Young, 2014).

Refining Phenomenality

Of the many treatments regarding the kinds of consciousness, few have been as influential in consciousness studies as access and phenomenal consciousness. Block (1995) is responsible for the claim that the concept of consciousness is not a cluster concept containing different kinds of relevantly similar concepts but rather a mongrel containing different kinds of fully dissociable states. The two kinds that Block is keen to distinguish and doubly dissociate from each other are access consciousness (A-consciousness) and phenomenal consciousness (P-consciousness; Block, 1993, 1995, 2001, 2007, 2008, 2009). However, the difference between these kinds of consciousness is definitionally opaque. Semantic and definitional clarity aside, a major difficulty with the distinction between A-consciousness and P-consciousness is that sometimes these states are differentiated and identified according to their representational content as an information-processing issue (Block, 1996, 2007, 2008), while at other times, P-conscious states are ostensively defined in light of their qualitative properties (Block, 1993, 1995). Furthermore, the difference between these kinds of consciousness has been challenged as conceptually ambiguous (Rosenthal, 2002, 2007, 2009, 2010) and incapable of scientific investigation (Kouider et al., 2012). Moreover, a review of the literature on olfaction suggests the distinction between these kinds of conscious states might not be applicable to olfaction because the experiential nature of A-consciousness and P-consciousness differs from the other modalities based on olfaction's unique neural architecture (Stevenson, 2009c).

While my distinction on offer could be encompassed within Block's framework of A-consciousness and P-consciousness, further refinements of his usage of P-consciousness would be required. As currently stated, qualitative consciousness and awareness provide greater precision and clarity in the demarcation of the relationship between these kinds of consciousness that is substantiated by experimental evidence from olfaction contrary to Stevenson's (2009c) claim. What will be offered is not meant to supplant but rather to supplement in a manner that can encompass the nature of olfactory consciousness.

Block's definition of phenomenal consciousness might be interpreted in one of two ways: (1) as referring to states that have a qualitative character of which the subject is conscious, although it might not be fully reportable or fully accessible; or (2) as referring to states that have a qualitative character, although the subject is in no way aware of being in the state (Rosenthal, 2002, 2007, 2009). The first interpretation corresponds to Nagel's (1974) concept of "what it is like" (WiiL), when the subject is aware of being in state *S*, and *S* has a qualitative character, although its content is not reportable or fully accessible. Nagel's precise usage of the phrase requires that there is a WiiL for the creature undergoing the experience. The notion of a phenomenal character of experience from a subjective point of view is inherent to the concept of WiiL. The latter interpretation of P-consciousness corresponds to qualitative consciousness, since the subject is unaware of being in state *S*, yet *S* has a qualitative character of experience. This latter kind arguably corresponds to that supported by Block's (2001, 2007, 2008, 2011) evidence for P-consciousness from subliminal vision and extinction studies.

Disambiguating these two kinds of phenomenality clarifies how the distinction of qualitative consciousness and awareness offers greater theoretical nuance in demonstrating that qualitative olfactory states can occur in the absence of subjective awareness. The fuller conception of phenomenality as a WiiL cannot be employed in providing empirical evidence for the dissociation of these kinds of consciousness as it smuggles in awareness. Assuming WiiL would muddy my claim that olfactory qualitative consciousness occurs in the absence of awareness, since some manner of subjective awareness is inherent to these states (section 6.3) and begs the question in demonstrating that olfactory awareness is always qualitative (section 6.4). Olfactory consciousness using the distinction between awareness and qualitative consciousness demonstrates what the original distinction was intended to capture. Qualitative consciousness does not smuggle in any aspect of awareness, and secondary-processing measures can establish the phenomenality of these states in the absence of awareness.

Methodologically employing a robust notion of awareness and, contrastively, the thinnest conception of phenomenal (qualitative) consciousness allows for greater conceptual clarity. Qualitative consciousness provides the starkest way of showing that phenomenality can occur without subjective awareness.[1] Furthermore, by stripping the subjective self-awareness aspect from qualitatively consciousness states, there should be no worry that some residue of subjective consciousness is smuggled in when it is shown that the olfactory states that we are aware of being in are always qualitatively conscious.

6.2 Measuring Olfactory Qualities

If our olfactory states employ a nonconceptual format, are not predominantly mediated by our visually dominated and linguistically mediated conceptual repertoire, and are not best accessed via semantic reports (or introspection), and if we are often unaware and inattentive to our smell experiences, then it becomes a vexing issue of how to measure if/when a subject is having an experience of a smell that carries an olfactory quality.[2] Given this difficulty, it has been argued that the best approach for conceiving of and measuring our perception of olfactory qualities is relative to a quality space, as it provides an independent means of assessing that an individual is perceiving an odorant as having an olfactory quality (Keller,

2017), which is independent of their subjective reports (Young et al., 2014). Quality-space theory (QST) provides the most viable way of measuring olfactory qualitative states and provides the impetus for further arguments that qualitative conscious states can occur in the absence of our conscious awareness of our smell experiences.

QST provides an alternative to the reliance on introspective reports drawn from subjective awareness. On this approach, mental qualities are measured based on their role in perceiving (Clark, 1993; Rosenthal, 2010). The core idea of QST rests on the discriminative function of perception. Perceiving involves discrimination of properties accessible by a particular sensory modality. Discriminating between stimuli with different odor properties requires that the subject is able to be in psychological states of two distinct types, each corresponding to the perceptible properties. Thus, the two types of perceptual state must differ in respect of some relevant sensory properties.

The conscious perceptual states that enable discrimination of sensory properties differ in respect of qualitative character. Identifying mental qualities in light of the different sensory properties that enable discrimination of perceptible properties allows the theory and its central method to claim that we can measure olfactory qualities and our experience thereof based on discrimination independent from conscious awareness. Perceptual states enable discrimination of sensory properties by differing in respect of mental quality. Mental qualities are the properties in virtue of which a creature can distinguish among the various sensory properties accessible to each perceptual modality.

For the purposes of smell, we can measure discriminative ability by testing for just-noticeable differences between discriminable sensory properties of two very similar odorants. Using JNDs, the quality space of smell can be constructed that represents all the discriminations that a particular individual can make among the sensory properties accessible using just the olfactory system. The quality space of smell provides both an exhaustive account of an individual's representational space of odors relative to their perceptual acuity (and genetics) and delimits their range of sensory transduction for individuating their sense of smell. Moreover, the space will provide not only identity conditions for the types of smells we can perceive but also a means of taxonomizing olfactory categories because mental qualities are the differential sensory properties of states that enable such discriminations. Hence, the very same space will also capture the differences and similarities among those qualities. Additionally, this measure provides a viable experimental means of comparing odor identity and individuation across individuals that is not mediated by subjective awareness, introspection, linguistic access, or background cultural associations—QST, with its proposed empirically reliable measure, can ascertain if and when an individual is having a qualitative experience, even in the absence of subjective conscious reports.[3]

The quality space of perceptible properties matches that of the mental qualities that enable discrimination among those sensory properties. One might incorrectly assert that such a match cannot be established without subjective awareness of the relevant mental qualities, and so QST cannot, after all, apply to nonconscious perceiving. But the matching relation is established by extrapolating from the space of sensory properties to that of mental qualities. That extrapolation is an inference about what enables the discriminations used to construct the quality space of perceptible properties and does not depend upon or follow from anything having to do with the subject being aware of their own states or their knowledge of the identity of the odorants transiting their nostrils. Moreover, the quality space of sensory properties need not reflect the chemical properties of the odorant, since the space is constructed not by appeal to the material nature of the stimuli but rather by how an individual discriminates among them.

Given that the quality-space framework does not determine the nature of olfactory mental qualities from the properties of the stimulus alone and does not presuppose that there are, in fact, sensory primitives for a modality, the dimensionality of the representational space will have to range over all possible individual odorants and olfactory mixtures relative to an individual olfactory acuity. With such a range of possible sensory perceptible properties, the dimensionality of our quality space for smells will vastly exceed that of the other modalities, given the estimates that we can smell up to one trillion different odors, and that just thinking in terms of organically generated odorants, the range is around forty billion. However, that is not to suggest that there might be an upper limit to the types of new and novel olfactory mixtures and synthetic compounds that might yield olfactory sensory qualities. Weiss et al. (2012) demonstrated that if the chemical structures within a mixture are similar enough and their molecular properties overlap, as well as controlling for intensity, then we can generate multiple instances of different mixtures with constituents ranging between thirty and sixty components that will have the same smell (as determined by subjective reports of the olfactory quality).

The reason for preferring this framework is that it provides an empirically viable and already established experimental method for determining an individual's representational space for olfactory qualities that fits with the methodological issues noted in attempting to study the nature of smell. Moreover, it is more parsimonious in its explanation of how our smell experiences can occur both with and without awareness but nevertheless always have a qualitative character that is implicated in mediating a wide range of our daily activities, despite not being aware of undergoing smell experiences. Quality-space theory thereby provides the necessary explanatory scaffolding for explaining how nonconscious qualitative olfactory states can have the causal influence they do.

A further strength of embracing QST as the foundational assumption in accounting for olfactory qualitative consciousness is that it provides a holistic representational mental space that can account for how genetic difference in ORN determination can generate differences in how we experience and classify smells, as well as the means of determining similarity of smell categories of olfactory qualities within and across cultures. The holistic quality space provides a means of blocking conceptual intuition pumps about an inverted spectrum and claims of epistemic access to a person's qualitative experience without proper experimental testing of their olfactory acuity. However, more importantly, it coheres with what has been argued previously: that our mental categories and perceptual states employ a rather unique format of compositionality. Thus, it is a viable means of accounting for how we represent mental qualities that preserves the FNCC format of smell in a comprehensive fashion, even for olfaction. What has been argued thus far is that the framework is viable for smell, coheres with everything else we have seen about how our olfactory system represents smells, and would allow the conservation of the format across perceptual, cognitive, and conscious states. However, no argument or evidence has been offered that we can have qualitative conscious states of smells in the absence of awareness, how these meditate behavior, or the relationship between qualitative consciousness and conscious awareness for smell. In what follows, the chapter progresses to show that all (or almost all if you prefer some hedging) olfactory mental states are qualitatively consciousness, such that merely eliciting cortical activation of initial perceptual/sensory states, even generating odor imagery, has a qualitative character. And while qualitative consciousness can occur in the absence of awareness, the converse is not the case.

6.3 Qualitative Conscious Smells without Awareness

Olfactory sensory states are such that they have a qualitative character, even if the subjects are unaware of being in them. Evidence that olfactory sensory states have a qualitatively character in the absence of awareness derives from research on blind smell, mate selection, the selection of social preference for social interaction and acquaintances, as well as the role of olfactory deficits in causing affective disorders. While none of these phenomena are decisive on their own, when taken together, they provide a host of evidence indicating that qualitative consciousness can arise independently of awareness.

6.3.1 Blind Smell

The existence of blindsight is a well-documented phenomenon that has played a role in shaping theories of consciousness. Blind smell, the olfactory analogue, is not nearly as well studied, but preliminary studies have shown that we can detect the presence of an odor in the absence of subjective awareness.

Schwartz studied blind smell using a detection task (Schwartz, 2000; Schwartz et al., 1994). Subjects had two vials placed in their hands: one containing an odorous solution, the other a non-odorous solution. They were instructed to sniff each vial and judge which one contained the odorous compound. In addition to monitoring their cortical activity using an electroencephalogram (EEG), the subjects were also required to report on the perceived concentration of the odor and their confidence in their own judgment. Initial results showed increased cortical activity on correct detection of the subliminal odor, which was correlated with the same neural activity as would be expected from the conscious presentation of an odor. The increase and similarity of neural activity might indicate that, even in the absence of conscious awareness, these subliminal presentations were conscious in some respect. The measures of subjective perception using concentration and confidence ratings corroborate these EEG results and establish the qualitative status of these states.

Given what was considered to be expected results, Schwartz (2000) inspected the previous data and discovered a large subgroup ($n = 32$) of more than a third of the subjects ($n = 86$) who had increased sensitivity. These sensitive subjects were able to detect the presence of an odor with 68 percent accuracy compared to the overall detection rate of 48 percent. In addition to their increased detection rate, they also displayed an increased confidence rating when they correctly detected the presence of the odor. Their self-reports, which were used to select the concentration necessary for the presentation of subliminal odors, demonstrate that they were unaware of the presence of an odor. Yet, their confidence rating on hits showed a direct relation, suggesting they were undergoing some manner of qualitative experience of which they were unaware. Based on these results, Schwartz concluded that the sensitive subjects possessed a level of consciousness, even when they were not conscious of undergoing this experience.

I am wary of attributing a level of conscious awareness to these states, but these findings somewhat show that olfactory sensory states have a qualitative character, even when subjects report no awareness of the odorant. The increased cortical activity on correct detection of the subliminal odor and the increased detection rate for the sensitive subjects do not on their own establish nonconscious qualitative olfactory states. Rather, the additional confidence ratings provide reason to think that these subjects had qualitatively conscious states in the absence of awareness. Since their confidence ratings overall were no higher than the normal and insensitive groups, what was of interest was that their confidence ratings increased in relation to their correctly detecting the odorant. Their confidence increased on hits, indicating that although they could not subjectively report the presence of an odorant, nonetheless there was something qualitative about their experience that allowed them to feel more confident in their judgment that the odor was present.

The sensitive subjects' subliminal perception of the presence of the odor affected their subjective feeling of confidence in their behavioral responses. Their confidence might be considered a qualitative character of subjective experience, fitting the definition of qualitative consciousness. There was something that it was like for the subjects to be in the state of correctly detecting the subliminal odor that was distinguishable from other perceptible states within the overall experimental task, and these states correctly modulated the subjects' behavior. Yet, such a conclusion needs the added proviso that, on my own framework and definition, the task was odor detection and not discrimination, such that, on my own terms, it is not possible to be certain that they underwent a qualitative experience of the olfactory quality of the odorant's smell.

Another study on blind smell involved the experimental design of a detection task requiring a judgment between an odor and no odor coupled with fMRI imaging (Sobel et al., 1999b). The subjects were instructed to sniff at given intervals and then report on the presence or absence of an odor (presented at two subliminal levels of odor concentration). Their results indicated that detection increased with the level of odor concentration, showing an increase in brain activity in those areas responsible for smell, even when the subjects denied undergoing any olfactory experience.

Sobel et al.'s (1999b) results are suggestive of there being dose-dependent unconscious olfactory processing, but the qualitative status of these states is dubious, since their detection task employed a subliminal odor without any further measure of subjective feedback. There is no doubt that there must have been something that it was like for the subject to be able to detect the presence of an odorant unconsciously, but this cannot be used to support a further claim that these states have a subjective qualitative character. Without further measures of the subject's experience of these stimuli, the claim that these nonconscious experiences contained perceptible properties that are qualitative is dubious. This is not to discount the findings of their experiment, but only to point out that further research on this phenomenon needs to be carried out.

Blind smell is an olfactory phenomenon reminiscent of aspects of blindsight that suggests olfactory states can occur in the absence of conscious awareness and subjective reportability. However, further research needs to be conducted to see the prevalence of sensitive subjects in the overall population, and further measures of subjective awareness need to be conducted in studies similar to Sobel et al.'s (1999b) to establish the qualitative character of these experiences for the subject in the absence of awareness.

6.3.2 Mate Selection

Evidence for the qualitative character of olfactory sensory states can also be gleaned from research on mate selection. Further research on human olfactory mate selection is required, but the initial data indicate that mate selection in humans is influenced by smell (Havlicek & Roberts, 2009).[4] We might not notice it, but our reason for choosing sexual partners might be that their immune system smells good to us.

Using olfactory cues, we select mates based on the synergy of our combined immune systems. If we mate with a partner whose major histocompatibility complex (MHC; alternatively termed "human leukocyte antigen" [HLA] in humans) is the converse of our own, this generates offspring with a more robust hybrid immune system. Thus, it is adaptive to be able to detect the structure of a possible mate's MHC.

However, the difficulty of studying human mate selection is readily apparent, given our inability to control for intervening variables. Most studies examining HLA mate choice have proven inconclusive, which could be attributed to these studies being conducted in heterogeneous populations in which the confounding effects of ethnic self-preference could not be controlled for. Nonetheless, the importance of smell in mate selection cannot be discounted. Based on questionaries rating the factors of mate selection, female subjects rated body odor as one of the most important factors in selecting sexual partners (Herz & Cahill, 1997; Herz & Inzlicht, 2002).

The qualitative character of a prospective mate's body odor plays a role in determining our choice of sexual partners, but to establish that this is related to odors derived from our HLA compounds, as detected by the olfactory system, and that this mediates actual mate selection requires three steps: (1) showing that humans can detect and discriminate the same MHC compounds that determine olfactory mate selection in rodents; (2) showing that we have the ability to detect the olfactory signature of HLA compounds and that these are treated as having a qualitative perceptible property; and (3) reviewing the literature of actual human mate selection in relation to HLA compatibility.

The causal mechanism for HLA detection is arguably the same as the mechanism responsible for MHC detection and recognition in animal models. Odors derived from MHC compounds play a role in determining mate selection in rodents. In mice and rats, it has been demonstrated that MHC recognition is accomplished by the olfactory system (Ehman & Scott, 2001; Yamazaki et al., 1979, 1980). Further research has also shown that mice, rats, and humans can smell the difference between the urinary scents of rodents derived from different MHC strains of mice (Beauchamp et al., 1985). Taken together, these studies show that mammals certainly employ MHC-based mate selection, and that the human olfactory system is sensitive to these same chemicals. When these findings regarding our olfactory sensitivity are combined with the research on human mate selection, strong evidence emerges that we engage in HLA-based mate selection as mediated by olfactory cues in the same manner as other mammals.

Not only do the odors derived from our HLA mediate mate selection, but it can also be shown that these odors have a qualitative character. Using men's two-day-old sweaty t-shirts, experimenters determined that females judged a t-shirt's odor as being most pleasant when it was derived from a man whose HLA system differed from their own (Jacob et al., 2002; Wedekind et al., 1995). In both of these studies, no single male body odor was universally agreed to be pleasant smelling—hedonic judgments differed across females relative to the dissimilarity of the donor's HLA. The major difference between these studies is that in Wedekind et al.'s (1995) study, the more dissimilar the HLA, the stronger the hedonic rating, while Jacob et al.'s (2002) results displayed a degree of HLA overlap in paternal lineage implicated in the hedonic rating of the sweaty odor. Nevertheless, both studies clearly implicate the olfactory system as a possible means for selecting mates based on the qualitative character of body odor as determined by HLA.

These positive results at best establish a correlation effect between the MHC of the donor and judged pleasantness. However, the work of Aksenov et al. (2012) demonstrated that MHC yields volatile odor compounds (VOCs) at the cellular level. Their study was the first to demonstrate that MHC compounds give off unique detectable odor signatures, such that a change in a single allele produces a unique odor fingerprint at the cellular level. The implication of these results is that each person's unique genetic makeup and, in particular, HLA complex will generate VOCs with a unique odor signature, thus allowing the connection between the judged hedonic profile of complementary HLA mates and the possibility that this is directly determined by the VOCs generated by a person's MHC compounds.

Further evidence that humans can detect the odor profile generated by the HLA complex can be found in studies of perfume selection. Pretheoretical intuitions suggest that the reason humans use perfume is to mask their body odor, since body odor on its own is commonly perceived as unpleasant. However, Milinski and Wedekind (2001) disproved the masking hypothesis by showing that we select perfumes that enhance our natural body odor. Not only is this effect only found for the self-selection of fragrances, which is explained by the fact that people usually purchase fragrances for themselves (Jellinek, 1951; Le Norcy, 1991), but also the judged pleasantness of an odor as correlated with body odor was consistent over a two-year period and not a matter of fluctuations in perfume fashion.

In addition to a perfume's enhancement of the pleasantness of perceived body odor, Lenochova et al. (2012) discovered that a self-selected perfume boosted the judged pleasantness of body odor relative to each person. This was borne out by their control group that presented a mixture of body odor and equally pleasant perfume that had not been selected by the subjects but which did not generate the same judged odor enhancement. However, it should be noted that their study was only conducted with male subjects. Because female body odor was not investigated, since it is generally less intense, making it more likely prone to a masking effect, further research was required.

A follow-up study by Milinski et al. (2013) used female subjects to address this concern and replicated the previous findings that we can select perfumes based on our own MHC profile but not that of others. Fragrances similar to the VOCs given off by one's own MHC have a boosting effect on body odor. Moreover, using fMRI imaging, Milinski et al.'s study revealed specific activation to peptides consistent with humans' ability to detect MHC-associated olfactory cues. Thus, HLA compounds generate VOCs with a qualitative character that we can detect and respond to behaviorally.

The strongest evidence that human mate selection preferences are driven by avoidance of those with HLA haplotypes identical to ours is derived from Ober et al.'s (1997) study of mate choice among Hutterites. Previous studies did not show an effect of HLA on mate selection but were conducted in heterogeneous populations where olfactory factors of mate selection might have been overridden by socioeconomic and ethnic factors. The Hutterite population served as a control because it is a small homogenous population with easily traceable genetic lineages. By looking at the HLA haplotype matches between spouses, they concluded that less of an overlap existed than would otherwise be expected if the selection processes were random. Ober et al. concluded that MHC-based mate choice is operant, even in humans. Furthermore, they suggest that the mechanism for HLA detection and structural comparison might be mediated by the olfactory system. The olfactory system is quite capable of such chemical structural analysis and comparison, as demonstrated by the aforementioned results that humans can detect and discriminate the relevant MHC odorants in rodents, are sensitive to MHC compounds of their own body odor, and judge body odors of complementary HLAs as more pleasant.

The Ober et al. (1997) study is by far the most significant source of data on the role of MHC in actual mate choice in humans because of its methodological soundness using a large sample within a closed homogenous population, thereby controlling for social and ethnic confounding factors. Of the studies on the role of MHC in mate choice, only four (to the best of the author's knowledge) have shown that MHC is significant in determining actual mate choice (Chaix et al., 2008; Giphart & D'Amaro, 1983; Ober et al., 1997; Rosenberg et al., 1983), while seven have shown no significance (Garver-Apgar et al., 2006; Ihara et al., 2000; Jin et al., 1995; Nordlander et al., 1983; Pollack et al., 1982; Sans et al., 1994). However, it should be noted that aside from the most recent study (Chaix et al., 2008), which showed a limited effect in only European American groups, the previous studies with positive results all used large sample sizes, thus controlling for the variegated properties of genetic variation as well as additional societal and normative practices in selecting mates. The null results of previous studies might simply be attributed to the lack of power due to small sample sizes in attempting to determine a complex human behavior with multiple intervening variables.

Additionally, Chaix et al. (2008) showed that MHC mate selection is apparent in European and American populations but not in African Yoruba populations. However, the statistical methods of testing their hypothesis were criticized not only because the significance could be attribute to extreme mate pairs within the groups but also for not correctly adjusting their statistical thresholds for multiple hypothesis testing (Derti et al., 2010). Nonetheless the critics agree that MHC based mate selection was an apparent, but not a robust result, which might simply be attributed to the small sample size.

Laurent and Chaix (2012) adjusted their previous results for multiple hypotheses to control for Derti et al.'s (2010) criticism and found that their results are still robust and significant. Moreover, they reiterated their claim that MHC might only be a determinant in some populations, while other factors such as ethnic background, inbreeding avoidance, and selection pressures from parasites might play a larger role. Therefore, further research is required. In reply, Derti and Roth (2013) maintain their original claim that the study is statistically flawed but yield to the multiple hypothesis issue and agree that lack of power and absence of evidence is not evidence that MHC does not play a role. Further research is certainly called for on the role of VOCs given off by MHC compounds in humans in the selection of mates across cultures. Currently, the evidence indicates that odorant detection of MHC compounds influences sexual mate selection, but the extent and mechanism require further study using more stringent and universal methodologies with large samples (Havlicek & Roberts, 2009).

The argument put forward in this section was that VOCs derived from HLA have perceptible properties with a qualitative character that are perceived using the olfactory system and modulate our mate selection behavior. Yet, we do not commonly attend to smell or its modulation of mate selection. While further research is required, at this initial stage, the evidence indicates that we select mates based upon the qualitative character of our olfactory states, even in the absence of awareness. We might simply love someone because their immune system smells nice. Human mate selection as determined by olfactory detection of HLA compatibility provides evidence for qualitatively conscious olfactory states, even in the absence of awareness.

6.3.3 Social Acquaintance Selection

Not only do consciously perceived smells modulate our mood and affective responses toward people (Herz & Schooler, 2002; Jacob et al., 2002), but subliminally pleasant and noxious odors also modulate our ratings of the likeability of social acquaintances (Li et al., 2007). Li et al.'s (2007) study showed that the hedonic value of an odorant subliminally modulates social preference. Using a simple odor-detection task (pleasant, unpleasant, neutral, and control) combined with a subjective rating of the likeability of pictures of faces, they demonstrated that pleasant and unpleasant odors presented subliminally had a physiological effect and modulated the subject's affective response toward pictures of human faces.

Independent of subjective awareness, there was a significant change in the heart rate of each subject relative to the valence of subliminal odors. Furthermore, unpleasant odorants caused the subject to rate the face as being less likable, while pleasant odorants had the opposite effect. The modulation of likability relative to odorant valence only occurred with subliminal odorants and quickly disappeared if the subject was aware of the smell.

Even in the absence of subjective awareness, we treat the odorant as having a given qualitative valence, which has a causal effect upon our attribution of properties to others. Arguably, this shows that qualitative consciousness is independent of our subjective awareness of the pleasant or unpleasant character of the smell. Even if one is unaware of undergoing an olfactory experience, the hedonic properties of subliminal odorants are implicated in social acquaintance selection. I might like you because you smell nice.

6.3.4 Anosmia: Argument from Absence

Olfactory pathologies provide evidence that it is not possible to have olfactory qualitative consciousness without olfactory sensory states. Moreover, they show that the robust qualitative character of our daily lives that smells impart, even though they go unnoticed, are taken for granted until they are gone.

Anosmia is the most common disorder of olfactory pathology in which individuals lose their sense of smell. In some cases, anosmia is due to the presence of a psychological disorder, but the vast majority of cases result from damage to the olfactory bulb due to either infection or head trauma. Individuals with fully functional olfactory systems modulate their sniffing in accordance with the pleasant or unpleasant character of an odor. Yet, anosmic individuals show no such response (Harland & Frank, 1997), demonstrating that the sniff response only occurs when the subject perceives the valence of the presented stimulus. Thus, anosmic individuals lack the ability to perceptually experience the qualitative character of olfactory valence.

In addition to their inability to perceive olfactory stimuli, anosmic individuals also experience a decrease in their hedonic quality of life (Miwa et al., 2001) and motivational anhedonia (Keller & Malaspina, 2013) that is often causally implicated in the further development of depression (Deems et al., 1991).[5] We are not aware of our olfactory experiences most of the time, but they imbue our lives with a qualitative character of experience, which is most striking in their absence.

To summarize, the argument from absence is that the absence of olfactory sensory states is causally implicated in lower quality-of-life scores and depression. Hence, these states are responsible for generating qualitative consciousness, even in the absence of awareness. The argument might not prove that all olfactory sensory states have a qualitative character, but the evidence certainly is significant and nicely fits with all the other mounting evidence thus far that olfactory states have a qualitative character, even in the absence of subjective awareness.

6.4 No Olfactory Awareness without Qualitative Consciousness

Qualitative olfactory consciousness dissociates from conscious awareness, but the converse is not the case, thus marking a further contrast within the framework under consideration regarding kinds of consciousness and, in particular, from Block's claimed double dissociation of access consciousness from phenomenal consciousness. The nature of our qualitatively conscious olfactory states surveyed above makes room for the possibility of qualitative states of which the subject is not consciously aware or, if you would like, nonconscious qualitative states. Thus, given that even subjectively unconscious states still have a thin qualitative character, it will be unsurprising that olfactory states that the subject is aware of will always have a qualitative character, which provides reason to think that, at least for smell, we might need to rethink the division between qualitative states and subjective (conscious) awareness.

Evidence for the claim that olfactory awareness is always qualitatively conscious might be derived from first-person reports and the reader's own awareness of olfactory experiences. Introspecting, remembering, or imagining an odor tokens some manner of qualitative olfactory experience. Just thinking about the smell of the fresh-cut grass elicits an olfactory experience for me. However, using first-person reports of phenomenology might be methodologically questionable. Aside from biasing us to consider only experiences that we are aware of as having a qualitative character, the veracity of olfactory first-person reports might be doubted, given our limited attention to olfactory experience and subsequent lack of awareness of our experience of odors (Sela & Sobel, 2010).

Veridical odor perception could establish that anytime we are aware of an olfactory experience, it has a qualitative character, but it is not a good test case. Situations of perceiving olfactory stimuli will activate a sensory state, which it has been previously argued are qualitatively conscious, thereby making one aware of an olfactory quality. Consequently, anytime we are aware of perceiving an odor, the conscious state has a qualitative character because qualitative sensory states are elicited as part of creating the perceptual state. Because first-person phenomenological reports are methodologically questionable and perceptual states might always have a qualitative character, olfactory imagery will serve as the test case for the conditional claim that if we are aware of an olfactory state, then it must be qualitatively conscious as well.

Methodologically, one could exhaustively search for a case in which we are perceptually aware of an odor but where the experience does not have any qualitative character. However, a stronger and more fatal test of my claim would be to find a state that does not elicit enervation of the olfactory sensory and perceptual states as caused by an odorant, that we commonly do not think would be qualitative, and that people find difficult to elicit in the first place (Herz, 2000) and check if these cases of olfactory awareness are qualitatively conscious. Mental states concerning olfactory experience are paradigmatic test cases of conscious awareness where we would not necessarily expect some level of qualitative character. Olfactory imagery provides exactly the right sort of test case, since it is a mental state concerning olfactory experience that can be about novel stimuli. Thus, what will be shown is that just volitionally thinking about smells elicits a qualitative character of experience.

While the phenomenon of olfactory imagery is primarily conceived of as an issue regarding the representational format of cognitive states in an analogous manner to visual imagery (Kosslyn, 2003; Kosslyn et al., 2003; Pylyshyn, 2003), it demonstrates that we can elicit a qualitative experience of a smell in the absence of an olfactory stimulus (reviewed in Rinck et al., 2009). Olfactory imagery demonstrates that all states of olfactory awareness are also qualitatively conscious. Experimentally, it has been shown that subjects can elicit the qualitative experience of smelling something in the absence of olfactory stimuli. Merely introspecting, imagining, or thinking about a smell elicits a qualitative experience of smelling an odor.

Even more fascinating is that olfactory imagery states mimic those of ordinary olfactory experiences such as odor mixing (Algom & Cain, 1991). Odor-mixing experiments yield the interesting results that when two similar odorants are combined to yield a configural compound, the resulting complex's odor is different from those of its constituent parts, while odorants that are dissimilar yield elemental compounds in which the odors of the constituents are clearly discernable. However, by simply changing the concentrations of the constituents, one can shift an elemental compound to a configural compound. What is of interest in olfactory imagery is that if one is asked to imagine the mixture of two odors and report the olfactory quality of the compound, the reports will mimic those given when smelling the actual odor.

However, for olfactory imagery to fully demonstrate that states of olfactory awareness are qualitative consciousness, it must be shown that these states' content and experiential properties are the same as the perceptual state of which it is a cognitive copy. The most obvious way to test for such an overlap of content and qualities would be based on self-reports as employed in the study above. Yet, these must be marginalized for the same reasons as introspective reports of past olfactory experiences—we simply cannot methodically test the veracity of subjective self-reports regarding olfactory imagery (Djordjevic et al., 2004).

Self-reports are doubtless invaluable tools, but they must be corroborated with other measures of the content and qualitative character. If olfactory imagery is to demonstrate that whenever we are aware of an olfactory experience, there is a qualitative character of experience, what needs to be shown is that these imaginary creations of an olfactory state have the same content and experiential properties as if the subject were perceiving the imagined stimulus.

Sniffing patterns are similar across both types of experiences, suggesting that to elicit an olfactory qualitative experience, one must manipulate the olfactory epithelium and bulb (the low-level sensory states), which then re-creates the experience by activating the olfactory cortex (Bensafi et al., 2007; Djordjevic et al., 2005, Rinck et al., 2009). To think about a smell, one must literally token the initial sensory and perceptual states, which are arguably qualitatively conscious.

However, a set of experiments by Tomiczek and Stevenson (2009) calls this into question and argues that the same perceptual state is not elicited. Rather, similar structures that are utilized for olfactory perception in general are activated. Tomiczek and Stevenson assert that we do not imagine a specific odor. Rather, there is a general overall increase in activation across areas in the olfactory system that are responsive to odorants similar to the imagined odor. While their results indicate that the imagined state does not have the same exact content and experiential qualitative properties, it does focus us in the right direction. Although these states might not be qualitatively identical, it might be possible to establish that the states perform the same role, and using secondary measures of how they accomplish these tasks, the best explanation of their content must involve qualitative character.

The difficulty is assessing whether the experiential properties are preserved from veridical odor perception through olfactory imagery. Since the veracity of subjective self-reports is difficult to ascertain, secondary-processing measures might be employed to verify that the qualitative character is conserved between olfactory imagery and perceptual states. Secondary processes are correlated properties or incidental effects (Cummins et al., 2001), such as speed, error rate, types of errors, fatigue, and so on, of the system when it performs a task. In addition to a state's performance of a role, there might be other secondary properties that can be used to evaluate whether the role was performed in the same way utilizing a very similar, if not identical, physical realization.

Secondary-processing measures are traditionally employed in debates regarding computational implementations of cognitive abilities. Yet, analogous measures are available in measuring perceptual states. In olfactory research, the perceptible property of valence (the perceived pleasant of unpleasant property of an odor) provides just such needed measures for assessing a state's qualitative property independent of subjective reports based on conscious awareness. Behavioral measures such as sniff rate and volume, response time, and heart rate can all be used as independent measures of perceived valence that indicate the olfactory system is treating these stimuli in the same fashion regardless of whether we consciously perceive the odors or can subjectively report the qualitative character.

Sniff rates relative to odor concentration and valence provide confirmation that not only can we be aware of an imagined odor, but these states also have a genuine qualitative character. Humans modulate their sniff rate and volume 150 milliseconds after the onset of a stimulus relative to its concentration and valence (Olofsson, 2014. The stimulus-dependent response of human sniffing is such that intense and unpleasant odorants are sniffed less vigorously and with a decreased volume. Measurement of olfactory motor responses to odorants is reliable enough to be used as a nonverbal measure of a human's detection and categorization of the odor (Frank et al., 2003). Additionally, as mentioned in section 6.3.4, anosmic individual show no such response, indicating that the sniff response only occurs in accordance with the subject experiencing the valence of the presented stimulus (Harland & Frank, 1997).

The methodology of verifying the qualitative character of an imaginary mental experience as being the same as veridical perception using measures of sniffing is currently employed in olfactory imagery studies. Using olfactory motor activity during imagery as a criterion to test the veracity of participants' claimed imagined olfactory percepts, Bensafi et al. (2003) confirmed that the same sniff parameters, including sniff volume, occur in imagery as in conscious veridical perception. They showed that not only is sniffing sensory dependent, but also sniffing in a similar fashion to veridical perception produces qualitatively more robust olfactory imagery (Bensafi et al., 2005; Bensafi & Rouby, 2007). Employing the same secondary processes increased the capacity for generating olfactory images and the strength of the olfactory quality, indicating that these subjects had olfactory experiences with qualitative character.

Kleemann et al. (2009) lent further support to the conservation of sniff rates as indicating the preservation of the same olfactory quality of experience and extended them to breathing patterns. The overall sniff volume and breathing amplitudes are the same between imaginary and perceptual olfactory states. Not only did subjects report an ability to imagine an odor in these experiments, but they also breathed and sniffed in the same fashion as if they actually perceived the odor. Moreover, preventing subjects from sniffing while imagining smells decreases the vividness of the imagined smell (Arshamian et al., 2008). These results further solidify the claim that olfactory imagery states are contentful mental states with qualitative character.

Sniff rate and volume are not the only secondary measures for assessing odor valence. Response time is faster in detection and discrimination tasks for unpleasant odors (Bensafi et al., 2003), and heart rate measurements show that we involuntarily categorize unpleasant odors (Bensafi et al., 2002). Given the role of sniffing in modulating olfactory imagery, it is unsurprising that olfactory imagery increases our detection rate of the target odor in a manner that is modality and content specific (Djordjevic et al., 2004, 2005). The subject's experience of odor valence during olfactory imagery can be verified using behavioral nonverbal measures such as sniff patterns and response time. These secondary measures establish the occurrence of the qualitative experience of valence in olfactory imagery.

The confluence of secondary measures of sniff rates (as well as other behavioral measures) enables the further inference that the sensory quality is being conserved in olfactory imagery. However, even with the corroborations of secondary measures, it might still be objected that the subjects are merely employing their tacit knowledge of olfactory perception in generating their reports and behavior during these experiments.

Similar criticisms have been used against visual imagery. Yet, critiques of this variety gain no traction in the case of olfactory imagery. The sniff responses in these cases seem to fly in the face of the idea that these states might be merely modulated by our propositional knowledge of olfactory perception but contain no actual qualitative character. It seems fanciful that we could modulate our breathing and sniffing patterns in such a precise and automatic manner when we barely even pay attention to these facets of our olfactory experience in normal cases of perception. Furthermore, the format of olfactory conscious experience is arguably not formatted and modulated by descriptive linguistic resources (chapter 4). If olfactory states across sensory, perceptual, and cognitive states are not compositional in accordance with the criitiques' prescribed propositional format, it would be rather surprising if the same format was not preserved in olfactory imagery. Additionally, it has been argued that the preservation of the FNCC format in olfaction is conserved even in olfactory imagery, as it explains the odor memory bump (Young, 2019c). Moreover, it has been shown that an increase in overall anhedonia yields a decreased ability for olfactory imagery (Bensafi & Rouby, 2007; Rouby et al., 2009), thereby implicating some level of qualitative character in mediating olfactory imagery.

We can elicit an olfactory experience in the absence of the odorous stimuli that has an olfactory quality mimicking veridical perception in terms of its subjective report, behavioral measures, physiological responses, and cortical activation. The fact that these states conserve and preserve all of these properties from veridical perception indicates that olfactory imagery states have a robust olfactory qualitative character. This therefore supports the claim that any time there is olfactory conscious awareness, these states are also qualitatively conscious. It also reaffirms one of the chapter's opening claims that our consciousness of smell provides reason to rethink the relationship between the different kinds of conscious states.

6.5 Conclusion

We are always qualitatively conscious of the odorants traversing our nostrils as smells. Yet, this is not meditated by introspective access and does not require linguistic mediation to ascertain. However, most importantly, subjective awareness with reportability is not the gold standard for ascertaining our consciousness of smells. Olfactory consciousness provides a more precise conception of the kinds of consciousness that are experimentally tractable. Moreover, when we take the time to explore the empirical literature on smell, it provides us with a rich theoretical starting point for rethinking the relationship between kinds of consciousness.

The evidence and arguments offered within this chapter establish that we can generate measures to ascertain that an individual is experiencing a robust qualitative perceptual state in the absence of subjective awareness. The method does not require linguistic access or verbal reports in keeping with olfactory naming abilities. It can be sensitive to perceptual acuity at the individual level while allowing for comparisons of mental quality spaces across cultures. More importantly, the holistic mental quality space is in keeping with the format of olfactory processing. And if all of that isn't enough, it provides the beginnings of a reply as to why we should not expect the phenomenology of smell experience to respect the linearity constraint developed against tracking intentionalist (Pautz, 2021) and odor theories such as MST (Skrzypulec, 2022) because the representational format of mental qualities will essentially not respect linear isomorphic mapping relations.

Olfactory qualitative perception can occur in the absence of awareness. Put in simpler terms fitting contemporary debates, we can smell unconsciously. The empirical evidence covered in this chapter is thus also relevant for the recent debate in philosophy of perception regarding the possibility of unconscious perception. Although I have not couched the chapter in such terms, all of the research covered in sections 6.3 and 6.4 can be used to establish that, at least for smell, there should be little doubt that we can perceive smells unconsciously when this is understood as qualitative perceptual states in the absence of awareness. As the conclusion is no place to make new arguments, I will simply note that the chapter could easily be adapted in favor of Block's position that unconscious perception is possible against Phillips's negative claim (Phillips & Block, 2016). In particular, it is not clear to me why we should agree with Phillips's foundational assumptions and definitional starting point in the debate that explicitly endorses Burge's (2010) conceptual framework of perception. Phillips's argument depends on unmotivated assumptions from Burge that generate an inflationist conception of perception as only being applicable to person-level states and not sub-personal levels of processing. How this distinction is meant to play out within the remit of empirically informed philosophy has always seemed dubious to me. It smacks of armchair theorizing without merit and seems theoretically unapplicable to smell perception, since introspective access, subjective reports, and linguistic mediation are only ill-conceived methods for theorizing about smell, especially in experimentally measuring the nature of our olfactory perceptual state. But even if we set aside these worries about personal-level approaches for demarcating the boundaries of perception, smell as a perceptual modality can meet even Burge's own criteria for being a perceptual system (Carvalho, 2014), such that, on Phillips's own terms, we should be allowed to conclude that, at least for smell, we can have unconscious perception.

Conscious subjective awareness is the liminal threshold for accessing smell experiences and thereby not the most valid or efficient access point as our initial starting point for theorizing about perception, mental representation, or our consciousness of smells. When all the differences about our sense of smell are accounted for and combined in the comprehensive framework that I am offering, then some interesting results emerge, suggesting that our ocular-centric foundational assumptions mediated by semantic conceptual analysis has generated some stinking philosophy. What will be suggested in the concluding chapter is that these implications have more far-ranging effects. Not only does olfactory philosophy provide reason to reassess entrenched philosophical debates, but it also calls into question dogmatic starting points within neuroscientific research and, in particular, empirical research on consciousness. Neglecting smell is not uniquely a philosophical enterprise—neuroscientific theories of consciousness were called out more than a decade ago (Young, 2012). Yet, the next chapter will show how their theories are still inadequate.

7 Stinking Theories of Consciousness

We may not always notice the world of odors enveloping us. Yet, they provide a powerful source of experience. Our sense of smell, the anatomical structure of the olfactory system, and its functional organization have profound consequences for the study of consciousness. While I have previously argued that the major scientific theories of consciousness are either false or inadequate as general theories because of their visuocentric methods and neglect of olfaction (Young, 2012), little has changed in a decade. The disproportionate dominance of vision is often simply taken for granted without even noting that experimental results must be relativized to visual consciousness. And while some are careful to note that they are developing a theory of visual consciousness, this is a rarity. For example, the entire debate couched in terms of the question "Is consciousness in the front or back of the brain?" (Boly et al., 2017; Koch et al., 2016; Odegaard et al., 2017; Storm et al., 2017)[1] only makes sense if vision is assumed to be the default modality that universally generalizes. The mere starting tacit assumption of the debate will exclude the nuances of olfactory processing regardless of the kind of consciousness under consideration. The focus of this chapter is thus to review how neglecting olfaction has negatively impacted neuroscientific theories of consciousness (including empirically tractable philosophical theories), such that a large portion of these theories are still either false or inadequate as general theories of consciousness.

A partial difficulty in studying consciousness is that everyone claims to know what it is. Yet, there is neither a consensus on what the exact phenomenon is nor agreement about the ideal methods for studying consciousness (Michel et al., 2019). We all claim to know what consciousness is from an expert subjective perspective, but what a subject of experience even is hides nuances of philosophical debates spanning millennia. Compounding this difficulty, theorists often start with background assumptions about the nature of consciousness and the methodology best suited for studying such a conception. Practically, this translates into the problem that each theorist's conception of consciousness should be noted before beginning an analysis of consciousness. Yet, this is often omitted.

Within this chapter, I will not add to this cacophony by imposing my own taxonomy that best captures smell from the last chapter upon other theories and models of consciousness. Going into this level of nuance will not be required in analyzing the applicability of the major neuroscientific theories of consciousness to smell. Rather, we can stick with the generally accepted claim that there are at least three different kinds of states that we commonly think of as consciousness: waking consciousness, conscious awareness, and phenomenal consciousness (for an in-depth introduction to the kinds of consciousness, see Berger, 2022, and for neuroscientific theories, see Mylopolous, 2022). Waking consciousness might be determined merely by physiological levels of arousal and is ascribed to a creature depending upon whether it is awake or asleep and responsive to its environment. More commonly, we conceive of consciousness as awareness, consciousness of, transitive consciousness, or access consciousness, such that the subject is aware of being conscious and their experiential content. Although entire theories are built upon the fine nuances between the definitions of such states, the common denominator is that we are sometimes aware of our experiences—we are aware that we are perceiving something or undergoing experiences.[2] The last notion of consciousness is that of phenomenal or qualitative consciousness: some (if not all) of our experiences have a qualitative character for the subject undergoing the experiences. To say that something is qualitatively conscious is to say no more than that there is some qualitative aspect for the organism to undergo the experience.[3,4]

With at least these three different conceptions of consciousness at play within the study of consciousness, this chapter will highlight the relevant kind for each theory. The focus will be upon the implications that our sense of smell brings to bear for the necessary conditions claimed by neuroscientific theories of consciousness. The general structure of each section is to state each theory's conception of consciousness, and its research strategy, methodological assumptions, evidence, and necessary conditions it requires for consciousness. After the overview of each theory, the consequences of neglecting smell will be set out. Possible replies on behalf of each theory will be offered, followed by a brief conclusion regarding the status of the claimed necessary conditions of consciousness, given the nature of smell. The theories of consciousness covered are: Merker's (2007) centrencephalic theory, Lamme's (2004, 2006a, 2006b) neurobiological theory, intermediate-level processing theories (IPT; including that by Jackendoff, 1987), Prinz's (2005, 2007) attended intermediate-level representations (AIR) theory, Mandik's (2000, 2005, 2009) allocentric–egocentric interface (AEI) theory, Crick and Koch's neurobiological specificity theory, the global workspace theories (GWT) of Baars and Dehaene, and Tononi and Edelmans's information integration theory (IIT).

7.1 Subcortical Consciousness: Merker's Centrencephalic Theory of Consciousness.

The centrencephalic theory (Merker, 2007) claims that consciousness can arise without the cortex and corticothalamic loops. Resuscitating Penfield and Jasper's idea that the midbrain reticular formation is supracortical in terms of function and control, Merker argues that the thalamus and corticothalamic relays are not required for consciousness. Merker is purposefully obtuse in defining consciousness. Since his theory concerns the necessary conditions of having experiences, which should be carefully contrasted with the necessary condition of being aware that one is undergoing an experience, it is most charitably treated as a theory of waking consciousness. His lack of definitional clarity is due to a very permissive notion of consciousness as whatever it is that makes experience possible, which is further clarified by his working assumption that the functional role of consciousness is to guide behavior, which can be facilitated in the absence of awareness.[5] Thus, Merker's methodology is to assume that the putative functional role of consciousness arose from the evolution of the visual system and our ability to navigate the environment to fulfill our homeostatic needs.

The driving assumption of the function of consciousness is apparent when he defines consciousness in accordance with the purpose it serves for the organism. It is in this vein that he claims that the brain evolved around the visual system, whose teleofunction is to select targets that will realize our homeostatic goals. Accordingly, the human cortex and thalamic connections evolved for the visual system. Yet, it is still possible to have target selection that does not involve the cortex. It is difficult to assess his claimed phylogenetic development of the brain without a complete evolutionary story regarding the development and selection pressures upon neuroanatomy and function. The assertion that the cortex's structure was evolutionarily sculpted in accordance with the development of the visual system seems empirically unmotivated and unsubstantiated. Despite justified skepticism in the assumption that the structure of the cortex was determined by the development of the visual system, it is worth noting that the specificity of brain areas for the visual system and the coercion of greater cortical tissue for vision are somewhat supported by research on the genetics of the olfactory system. Buck and Axel's (1991) Nobel studies regarding the genetic basis of olfactory receptors have shown that mice have one thousand genes for producing olfactory receptors, only a portion of which generate olfactory receptors. Depending on the species, the number of pseudogenes could range from 10 to 60 percent. One of the interesting findings using comparative genetics in animal psychology is that species with trichromatic vision have a proportionately higher percentage of pseudogenes (Gilad et al., 2004). The hypothesized explanation of this is that as visual acuity increases, the utility of olfactory acuity decreases. Alternatively, olfactory acuity may be sacrificed for vision, given constraints on the amount of tissue available to the organism based on caloric consumption and space. Yet, it should be noted that alternative interpretations of the evolutionary function of olfaction suggest that olfactory bulb volume might be selected precisely for environmental navigation (for a brief review, see Young et al., 2020).

Another contentious aspect of the centrencephalic proposal is that the visual system is primarily responsible for object tracking and goal setting, in light of action planning, to acquire the object of desire. There is evidence of our human olfactory system's capacity to track objects across distances, depending upon the concentration level and the presentation to the different nostrils at different times (J. Porter et al., 2005, 2007). Moreover, a review of olfactory navigation using odor plumes across species suggests that the system is adept at identifying odor targets and navigating through a complex environment to the odorants source (Young et al., 2020). When our olfactory tracking abilities are combined with olfaction's role in detecting social relations (predator, family member), food sources, and possible mates, Merker's claimed centrality of the visual system seems dubious.

Setting aside these questionable underlying assumptions, Merker does provide contemporary evidence for Penfield and Jasper's claim regarding cortical control that the midbrain reticular formation was supracortical in terms of both function and control. Evidence for their claim derives from surgical procedures performed on patients with epilepsy in which the removal of brain tissue was required to alleviate their symptoms. Based on their work with 750 patients who were awake and responsive during the surgical treatment, so as to ensure that the key areas underlying conscious control and cognition were not affected, Penfield and Jasper were impressed by how much cortical tissue could be removed without a patient losing consciousness. Their neurosurgical findings provide support for the centrencephalic conclusion that the cortex is not necessary for consciousness or cognitive activity generally, but rather that the midbrain and its extension (including the nonspecific thalamus encompassing the midline, intralaminar, and reticular nuclei) are responsible for cognition.[6] According to this incarnation of the centrencephalic proposal, while the cortex is not necessary for all forms of cognition and consciousness, some degree of cortical mediation is still required. Merker's own stance on the requirement of cortical mediation is explicit that consciousness itself need not be mediated by any cortical connections and can occur independently in the midbrain. To make this audacious move and the original theory even more minimal, Penfield and Jasper's approach is updated using Thompson's (1993) subcortical general learning system, the Sprague effect, midbrain target selection, and anencephalic children.

Thompson's (1993) subcortical general learning system contradicts our common thinking that the cortex and conscious awareness are required to learn about and navigate our environment. Thompson showed that animals are able to learn even with substantive cortical lesions. Subcortical areas might be considered more important than cortical areas when considering a system-level analysis of neural function. Thus, his research provides further support for Penfield and Jasper's proposal that "certain upper brainstem systems in receipt of convergent cortical projections occupy a superordinate position" (Merker, 2007, p. 66).

Continuing the theme of considering the midbrain to be supracortical, the Sprague effect concerns individuals with cortical damage who have lost some visual abilities, but upon further damage to the connections between cortical areas and subcortical tissue, several visual abilities reemerge. Their restored visual capacity is limited to the ability to orient and approach locations of moving visual stimuli in space, and visual pattern discrimination does not recover after midbrain intervention. The Sprague effect supports Merker's claim that just thinking in terms of cortical deficiencies inflates the functional necessity of the cortex. Cortical deficiencies on their own need not necessitate deficiencies of consciousness, but the cortex's connection to neighboring areas explains how cortical deficiencies impact upon midbrain proficiencies that are responsible for consciousness.

Further support for the centrencephalic proposal can be gleaned from our abilities for target selection, goal setting, and action planning, which demonstrate the supracortical function of the colliculus in the control process (Carello & Krauzlis, 2004; McPeek & Keller, 2002). Furthermore, the colliculus sums up and decides, from the possible actions available to an organism, which to execute (Allport, 1987; Brooks, 1994; Dean et al., 1989; Isa & Kobayashi, 2004; McFarland & Sibly, 1975). Evidence for the colliculus performing a functionally higher role than the cortex in target selection is supported by collicular lesion studies with macaques that showed a great deal of compromise in their sophistication and scope of target selection (Albano & Wurtz, 1978; Casagrande & Diamond, 1974; Denny-Brown, 1962; Mort et al., 1980; Schiller et al., 1979; Schiller & Lee, 1994; Schneider, 1967).

The most fascinating part of Merker's theory is his discussion of children with the medical condition of anencephaly: "Anencephaly is the medical term for a condition in which the cerebral hemispheres either fail to develop for genetic developmental reasons or are massively compromised by trauma of a physical, vascular, toxic, hypoxic-ischemic, or infectious nature at some stage of their development" (Merker, 2007, p. 78). Most cases of anencephaly occur in children who are missing the vast majority of their cortex. According to medical practitioners, these children fall within the definition of being brain-dead. However, observations of them in home settings indicate that they have experiences that seem to have a qualitative character.[7] They might not possess awareness of their experiences, but they appear to have preferences for different kinds of experiences. What he suggests is that these individuals have qualitative consciousness without having a cortex—that is, they have experiences of some nature, but they do not possess an awareness of these experiences. Thus, anencephaly provides reason to conclude that a precondition of consciousness need not be the cortex or corticothalamic loops.

The problem with using these cases as criteria of consciousness is not that they might teach us about the sufficient underlying neural conditions that are a prerequisite for consciousness, but rather it is to do with the nature of consciousness that these children are claimed to have. It is quite clear that they attain waking consciousness and do not possess awareness, but Merker's evidence for their attainment of qualitative consciousness is far from demonstrative. A critical flaw of the centrencephalic proposal is that it is unclear whether it is a theory of consciousness or merely a theory of the preconditions necessary for waking consciousness.

Anencephaly and our subcortical learning system support the midbrain as a precondition for waking consciousness, but few would doubt that the midbrain, including the reticular formation of the thalamus, is necessary for waking consciousness. Lesion studies have shown that damaging the reticular formation of the thalamus causes a lack of waking consciousness (Baars, 1988, 1997). So, what is Merker adding that is novel? His greatest contribution to the study of consciousness is the description of anencephalic children. Anencephaly suggests the possibility that individuals may have phenomenal consciousness or qualitative aspects of experience even in the absence of awareness, such that awareness is neither necessary for nor identical to qualitative consciousness.

But what about olfaction? One of the key findings of the exposition of the anatomical connections of the olfactory system is that our experience of odors does not require the thalamus or corticothalamic loops, since the neural connections project directly from the olfactory bulb to the cortex.[8] The unique anatomy of the olfactory system provides an immediate counterexample to Merker's entire proposal that consciousness need not involve the cortex and corticothalamic loops. While the last half of his claim—that corticothalamic loops are not necessary for consciousness—is true,[9] the olfactory system's direct projection to the cortex without thalamic connections suggests that, at least for olfaction, the cortex is required for conscious awareness, which highlights the importance of tracking a theory's target type of consciousness and supporting evidence. The centrencephalic theory of consciousness might be adequate in providing some insight into the neuroanatomy of waking consciousness, but it will certainly not be able to account for conscious awareness, given olfaction's unique cortical projections.

Merker's approach is incorrect as a general theory of all types of consciousness. Yet, it provides a springboard for launching an attack upon the overvaluation of the cortex and the stranglehold that the notions of access consciousness and transitive consciousness have upon theorizing about consciousness. Prevailing orthodoxy is to begin by explaining the datum of conscious awareness from which an understanding of qualitative consciousness is then derived. Merker's approach sets out the initial conditions to argue for a bottom-up approach in which waking consciousness is a necessary condition for phenomenal consciousness, which then forms a constitutive condition for awareness.

7.2 Lamme's Neurobiological Theory of Consciousness

Lamme (2004, 2006a, 2006b) seeks to explain the nature of conscious experience in terms of neuroscience by arguing that the only way to define consciousness successfully is by using neuroscientific approaches and concepts. He thinks we can circumvent the problems of reportability by changing the conceptual framework that we use to talk about our experiences. Commonly, we think of conscious states as those that we can report being in. However, Lamme counters that using reportability as the sole criterion for consciousness has the consequence that the entire right hemisphere of the cortex could not be considered part of the neural correlates of consciousness (NCC). Language and reportability cannot be the sole criterion for ascertaining if a state is conscious unless we disallow the entire right hemisphere from having consciousness merely because it does not have access to language centers.

Lamme distinguishes between awareness and attention in the visual system, scientifically cleaving the two cognitive functions from each other, so as to allow a conscious experience of a visual entity, even when we are not attending to the object.[10] He claims that we can neuroscientifically separate attention from awareness, such that attention is responsible for reportability, yet we might be aware of far more states than we can report. While the separation of these two cognitive faculties or functions might be suggestive of Lamme's approach falling within the parameters of conscious awareness, it is quite clear that he identifies awareness with access consciousness and states that are conscious but not attended to with Block's notion of phenomenal consciousness.[11]

Block (1996) famously argues that the concept of consciousness is not a cluster concept containing lots of different kinds of relevantly similar concepts, but rather a mongrel containing different kinds of states that nonetheless share the same term. The variegated nature of consciousness allows Block (1993, 1996, 2001, 2007, 2008, 2009) to distinguish between P-consciousness and A-consciousness. The most charitable interpretation of this distinction that fits with Lamme's approach is that P-consciousness occurs when there is some information that is not available for reasoning, reporting, or rationally guiding action, but there is still something that it is like for the subject undergoing the experience. By contrast, A-consciousness occurs when the representational content of a mental state can be used to make inferences within rational thought processes that are reportable.

Lamme's (2003) background assumption is that consciousness is not functionally realized in a manner that is localizable, as opposed to Crick and Koch who think we can find one particular area or region of the brain that is responsible for conscious awareness.[12] Furthermore, consciousness must be realized based upon recurrent processes as opposed to feed-forward systems (Lamme, 2004, 2006b). The evidence for his theory derives from change blindness, inattentional blindness, backward masking, and transcranial magnetic stimulation (TMS) studies (Lamme, 2006a). Change blindness (Lamme, 2006a) and TMS studies (Lamme, 2003), he claims, indicate that it is possible for us to be aware of something, even though we do not or cannot attend to it.

Backward masking and TMS studies, Lamme argues, provide reason to doubt the adequacy of feedforward networks for conscious activity. Rather, FFS activation combined with recurrent cortical loops provide necessary and sufficient conditions for consciousness. So long as the current connections occur within parts of the cortex that are not accessible to higher cortical areas such as memory or language production, reportability is not possible, but these states could still be phenomenally conscious. Since visual awareness requires recurrent processes for consciousness, he infers that a similar structure of recurring connections must underlie phenomenal experience as well.

Lamme's theory is best analyzed in terms of two separate but constitutive claims. First, access consciousness occurs when recurrent connections sweep from the back of the brain in the visual system to the prefrontal and frontal areas of the brain. Second, recurrent connections within the extrastriate areas of the primary visual cortex are phenomenally conscious, even though they are not projected to frontal areas and cannot be reported linguistically. Although both kinds of consciousness share the same underlying theoretical assumption, such that a necessary condition of consciousness is that there are recurrent connective sweeps, the requisite conditions for attaining each kind of consciousness are slightly different.[13]

Lamme is wonderfully moderate in only claiming to have a theory of visual consciousness, such that its applicability to our sense of smell might be limited. However, we can still worry about the generalizability of the overarching claims that he has regarding the nature of consciousness. Attended conscious states are those states that I am conscious of and are accessible to me based upon the role of attention. But the role of attention is not clear when it comes to olfaction, and we do not attend to most of the incoming olfactory stimuli, such that some have gone so far as to suggest olfactory perception is a constant state of blindsight (Castro & Seeley, 2014; Sela & Sobel, 2010). Yet, the vast majority of olfactory stimuli have conscious effects.

Lamme asserts that attention is required for the formation of long-term memories because mental states, which are not attended to, cannot attain long-term memory storage and be reportable. The role of attention in memory formation might be questionable based upon the olfactory system's direct projection to the hippocampus, which is traditionally implicated as the area responsible for memory formation, consolidation, and storage. Although these anatomical differences do not falsify his approach, olfaction's direct projection to areas of the brain responsible for memory creates a problem for his differentiation of A-consciousness from P-consciousness, given that olfactory states might project directly to the hippocampus and limbic areas in a manner that is far faster than the projection to the cortex. Moreover, imaging studies on human olfactory capacities show that olfaction might have a dedicate hub for odor memories not subserved by linguist naming abilities within the primary olfactory cortex (Zelano et al., 2009). Given these anatomical and functional differences with the workings of smell memories, this aspect of his approach requires further scrutiny.[14]

Lamme's theory requires further elaboration concerning what is causally responsible for generating attentional mechanisms. Within the visual system, the salience of the object itself might cause us to attend to the object. However, with regard to the olfactory system, questions arise about which aspect of an olfactory object should be considered its salience and how it is identified. The most likely candidate is the level of concentration. However, concentration on its own will not be sufficient because the activation pattern that will result after increasing the concentration will amplify general overall olfactory receptor neuron and glomeruli activation, which can yield a change of the percept itself.[15] Odors are experienced in a different manner, depending upon the level of concentration presented to the olfactory system. Another option might be to identify salience in terms of the valence of the odor, since this property of a smell has been shown to guide approach and avoidance behavior directly. However, as argued in chapter 2, odor object identity is best conceived of in terms of olfactory quality and not necessarily valence or hedonics. Thus, the issue in adapting his theory to olfaction would be identifying the correct analogue of a smell's salience that maps onto his ocular-centric approach or developing a theory of smell perception as modulated by hedonic attentional mechanisms, which sounds interestingly promising but well beyond the scope of my research.

Visual awareness and attention might require recurrent connections and loops between frontal and visual areas of the cortex, while recurrent processes within the visual system itself are those Lamme identifies with P-conscious states. According to this line of thought, the more pronounced the object becomes in terms of its lines, edges, shading, and color, the more phenomenal the states become. As such, P-consciousness involves the intermediate level of processing above area V1 in the visual cortex. The implication this has is that the object is fully bounded from the initial sensory level, which would imply the possibility that P-consciousness could arise in the olfactory bulb, which is certainly possible but seemingly implausible.

As currently proposed, the olfactory system and our conscious experiences of smells create difficulties for Lamme's theory, but his theory is adaptable enough to encompass the aforementioned problems. Some adjustments to the constituent parts of the theory might be required for its application to the olfactory system. The only aspect of the approach that is truly questionable in application to smell involves the intermediate levels of processing being phenomenally conscious. Since the existence of intermediate levels of processing within the olfactory system is an issue that arises in the next section concerning the levels of olfactory processing in intermediate-level processing theories of consciousness, let's now turn our attention to them.

7.3 Intermediate-Level Processing Theories of Consciousness

IPTs of consciousness all concern the nature of conscious awareness and share the claim that consciousness arises at an intermediate level of cognitive processing. Jackendoff's (1987) IPT, Prinz's (2000, 2005, 2007) attended intermediate-level representations theory, and Mandik's (2000, 2005, 2009) allocentric–egocentric interface theory all agree on where consciousness occurs within the stream of cognitive representations yet disagree on the necessary conditions for consciousness or the nature of the representational format. Jackendoff presents the initial form of the theory from a computational representational standpoint but does not provide neuroscientific evidence for the theory. Neuroscientific evidence for the IPT is later added by Prinz's AIR theory, with the added stipulation that only attentively modulated intermediate representations become conscious. Mandik's AEI theory departs from the previous theories by elaborating the perspectival requirement as its theoretical starting point.

Since IPT theories share theoretical underpinnings, evidence, and methodology, each will be summarized individually, but their shortcomings will be handled together. My argument will not be that these theories cannot explain our consciousness of smell, as empirically the matter is unclear. Rather, as things currently stand, some of their necessary conditions of consciousness are not applicable to smell. Where possible, I will suggest how to adapt the theories to accommodate olfaction, but my general prognosis is that these theories need to rethink their requirements for being an intermediate representation.

7.3.1 Jackendoff's Intermediate-Level Processing Theory

According to Jackendoff's IPT, consciousness arises at an intermediate level of computational processing. Borrowing heavily from Marr's (1982) work on visual processing, he suggests that within the visual system, consciousness arises at the level of the 2½-D sketch, while our experience of language occurs at the level of formal logical processing for language and the level of notes for the surface of our music awareness. Jackendoff's theory proceeds in two steps: first, the levels of processing are outlined (how cognition arises at these three levels is explained in terms of vision, language, and music comprehension), and then our phenomenology of awareness in the different modalities is used to fix the point at which consciousness occurs within the hierarchy of processing. The inference is from the character of the representations of our introspective awareness to the level of information processing that best fits these representational characteristics.

His methodological assumption is that a computational series can generate the same distinctions within the computational processes that are found within our awareness of our experiences. He does not think that this methodology answers the mind–mind problem or how computational processes lead to qualitative states. This makes the notion of consciousness at play within Jackendoff's theory that of conscious awareness.

Jackendoff's theory proceeds from a computational perspective to predict the level of computational processing at which consciousness occurs. To ascertain the level of representational structure at which consciousness occurs, he offers four prima facie possibilities: first, that consciousness arises at the sensory level of processing; second, that consciousness happens at an intermediate stage between the lower and upper levels of processing; third, that consciousness arises at the central level of processing, such that there is some discrete central processing unit that gathers the representational outputs from across multiple modalities and outputs them in an unified code to higher-level representational processing[16]; and fourth, that consciousness only arises at the highest level of computational processing, which involves full conceptual structure.

Having stated these possibilities, Jackendoff argues that only the intermediate level of processing fully captures the phenomenology of our experiences. While his explanation of consciousness is meant to derive from the computational processing of mental content, our reports of the content of our awareness in different modalities act as the decisive factor and evidence for his approach. In accounting for our awareness of the content of mental states, phenomenological reports are employed to garner evidence for the conclusion that what we are aware of is not the conceptualized level of cognitive processing or the representations at the lower sensory level but rather representations that arise at the intermediate level, which involve a vantage point of an individual as located in space undergoing the experience. Evidence for the intermediate stage being the correct level is offered regarding the nature of our linguistic, auditory, and visual experiences.[17] However, what little he does say about our olfactory experience is simply not true.

Jackendoff asserts that taste and smell are so experientially intertwined that one cannot distinguish between the two based upon the mere content of our phenomenal experiences. Taken at face value, it is unclear whether he thinks we cannot or that we do not. Since taste and smell are often conflated, we do not distinguish between these modalities a good deal of the time. However, that does not demonstrate that we cannot distinguish between the two once we attain a better grasp of the nature of smell, as shown in chapter 3 and argued in Young (2023) that we can distinguish our sense of smell from flavor using the object of perception and perceived sensory qualities.

According to IPT, attention is not necessary for consciousness, but it plays a definitive role in focusing our awareness of the elements of our attended experience. Attention modulates the amount of detail of the constituent structure of representations that gets encoded at the intermediate level. Yet, this claim is extremely questionable (if not false), assuming my argument in chapter 4 is correct[18] that odors occur in a combinatorial but not concatenatively compositional representation format. Additionally, our inability to identify more than three or four components of an olfactory mixture nicely shows that top-down attentive mechanisms do not enhance the details of the constituent structure of compositional olfactory representations.

An alternative interpretation of Jackendoff's claim as applied to olfaction might be that the constituent structure of an odor experience together with its valence, memories of past occurrences, its associations, and emotional significance are modulated by attention. Yet, even this interpretation is questionable based on both our general lack of attending to smell experiences (Sela & Sobel, 2010) and the arguments in Young (2011, chapter 3) that the content of olfactory experience outruns our ability to represent its constituent structure conceptually, as well as the findings from chapter 5 that olfactory cognitive categories are structured in a holistic fashion to accommodate a range of similar yet nonidentical olfactory objects. Thus, the formative nature of olfactory representation is not compatible with the basic tenets of Jackendoff's approach concerning attention. Furthermore, if olfactory processing is combinatorial but not classically compositional, then Jackendoff's entire methodology of analyzing distinctions within the phenomenology of our awareness in terms of computational constituent structure distinctions is called into question.

To summarize, Jackendoff claims that consciousness arises at the intermediate level of cognitive representational processing relative to each modality and that the general characteristics of these representations are perspectival, modality specific, and not fully conceptualized.

7.3.2 Prinz's Attended Intermediate-Level Representation Theory

Prinz's (2000, 2005, 2007, 2012) AIR theory of consciousness is modeled upon Jackendoff's, provides neuroscientific and psychological evidence to support its phenomenological argument, and gives a more substantive role to attention. According to AIR theory, consciousness arises at an intermediate level of processing when we attend to those states. Borrowing from Jackendoff's use of Marr's theory of vision, Prinz's arguments primarily involve the visual system and the mechanisms of visual processing. Although he admits that the details of Marr's model are outdated, Prinz maintains that the general idea of the three levels of processing is adequate in spirit to capture the levels of processing within our cognitive architecture. The first level is the primal sketch, which merely encodes the sensory constituents of an experience. Within the visual system, the primal sketch represents nothing more than the lines and edges of an object independent of each other. At this level, the smallest representations of the system occur. Intermediate-level processing occurs from an individual's perspective and represents objects with properties, neither of which have yet been fully abstracted from the particularity of the percept. The higher-level processes are responsible for representational content that is categorical, at an abstract level of conceptualization, which does not represent the information from a vantage point.

Prinz echoes Jackendoff's methodology by attempting to identify the level of processing that corresponds to our phenomenological reports of what we attentively claim to be conscious of. The inference is from the characteristics of the mental representations that we are aware of to the level of representation within the cognitive hierarchy at which this occurs. Methodology in hand, his arguments reflect those of Jackendoff's, such that we do not experience the sensory-level representations and are not conscious of our experiences in an abstract manner. Since our conscious experiences are from a vantage point, conscious experience must arise at the intermediate level of processing. Evidence for these claims is further supported by neuropsychological data meant to confirm that consciousness does not arise in the primary visual cortex but rather at the extrastriate area of the visual cortex.

Having ruled out the lower sensory levels, Prinz argues that consciousness cannot occur at the higher levels because representations at these levels of processing are too abstract to match the format of our reports of conscious experience. While consciousness might not arise at these high levels, our conscious experience is certainly modulated by them—a role that Prinz attributes to attentional selection. The essential role of attentional selection in consciousness sets the AIR theory apart from IPT, since according to the former, it is a necessary condition for consciousness, while for the latter, it only modulates it.

7.3.3 Mandik's Allocentric–Egocentric Interface Theory of Consciousness

Mandik's (2000, 2005, 2009) AEI theory states that consciousness arises at a stage of computational processing between the personal and conceptual domains of representation. The distinction between these two different kinds of representations derives from the notion of a perspective as required for planning guided motor actions. With this starting point, AEI theory clarifies the perspectival requirement that distinguishes high-level and intermediate-level representations. Using pictorial representations as a launching point, Mandik argues that all representations are encoded either from a standpoint or from a third-person perspective.

Allocentric representations are representations without a perspective, while egocentric representations derive from a perspective or vantage point. AEI theory states that consciousness arises with recurrent loops between the allocentric and egocentric representations, since the sensory level of incoming stimuli cannot be conscious and higher-level cognitive states may only mediate what is going on at the intermediate stages. The theory is best interpreted as claiming not that consciousness is localized at an intermediate stage, but rather that it occurs as recurrent processing loops between allocentric and egocentric levels of processing. Conscious states are a hybrid of allocentric and egocentric representations. They are not intermediate-level representations but rather an amalgamation of these two types of representations. While consciousness arises between these two levels of processing, this does not necessitate the existence of an intermediary level of processing.

Given the similarities between all three theories, it is not surprising that their evidence derives from the same experimental studies. Mandik's arguments and evidence are nearly identical to those of Prinz, who in turn echoes

Jackendoff. He argues that we are not conscious of all sensory states because we are not at all aware of the retinotopic mapping of stimuli or representations within the lateral geniculate nucleus. Furthermore, he argues that what we are conscious of is not the same as the representations that are found in area V1 of the visual cortex. Additionally, we are not aware of full allocentric visual representations from the frontal cortex or hippocampus. His evidence against the claim that low-level sensory states can be conscious is more or less identical to that cited by Prinz, while the arguments against the claim that high-level processes can be conscious are supported by appeal to the cases of blindsight and motion-induced blindsight.

While Mandik agrees that we are conscious of intermediate-level representations, his conception of a mental representation and his theory of levels differs from both the IPT and AIR theories. For this reason, in the next section, I compare and contrast the three theories.

7.3.4 Summary: General Requirements of All Intermediate-Level Processing Theories and Differences

According to all three of the theories, we are conscious of intermediate representations from a perspective or vantage point, such that they are neither abstractions nor conceptualizations of our experiences. Although they all agree on what we are conscious of, they do not agree about how consciousness arises, whether it be merely intermediate representations, intermediate representations modulated by an attentional mechanism, or mongrel representations formed by loops between top-down and bottom-up processes.

Jackendoff, with his IPT theory, claims that attention attenuates the detail of the constituent structure of the intermediate-level representation but does not consider it necessary for consciousness. By contrast, Prinz argues that attention is required for these representations to become conscious. AIR theory posits attentional mechanisms, which select the intermediate representations that will become conscious. While being an intermediate-level representation, according to Jackendoff, might be sufficient for being conscious, Prinz maintains that attending to these representations is also necessary. Mandik's AEI theory does not concern the role of attention. Rather, he claims that consciousness arises with recurrent connections between the egocentric sensory level and the allocentric conceptual level.

A further difference is Mandik's conceptualist framework according to which one's conceptual repertoire determines the experience one can have. Jackendoff and Prinz allow that incoming stimuli and top-down processes may modulate our conscious experience, but they do not think they are part of the essential conditions for having an experience. According to the AEI theory, a necessary condition for attending to the sensual world is that there should be recurrent connections between conceptual levels and egocentric sensory levels.

7.3.5 Criticisms of Intermediate-Level Processing Theories

Rather than question the adequacy of the evidence provided by Jackendoff, Prinz, and Mandik, in this section, I raise doubts about IPT's applicability to the case of olfaction based on phenomenological concerns, empirical studies regarding olfactory imagery, and concerns derived from IPT's background assumptions regarding representational cognitive processing, hierarchy, and formative nature. The main worry is that the three levels of computational processing that IPT posits are not clearly seen within olfactory processing. IPT may be perfectly adequate as a theory of vision and of the other sense modalities, but the scientific evidence and arguments upon which they are constructed do not generalize to our sense of smell.

The main challenge IPT faces is the applicability of their criteria for being an intermediate level of representation. Indeed, the criteria that Jackendoff and Prinz developed (and which, for the most part, were endorsed by Mandik) derive from Marr's computational model of vision processing. An intermediate-level representation is minimally a representation that is reportable based upon one's awareness of the state; arises from lower-level perceptual states; is modulated, attenuated, or selected by higher-level processes; and is derived from a vantage point (relative to a perspective). Even if it is possible for Jackendoff, Prinz, or Mandik to show that these levels of processing occur within the olfactory system, the suitability of applying IPT to olfaction would still be challenged if the intermediate-level representations in olfaction are not formatted in the representational manner they assume.

7.3.5.1 Anatomical and functional hierarchy In chapter 1, the key anatomical and functional differences between the olfactory system and the systems of the other modalities were explained and explicated. These differences create the unique nature of stimuli transduction and processing within the olfactory system. The first line of criticism concerns whether the hierarchical structure posited by IPT can be accommodated by olfactory processing.

Based on the claimed existence of a hierarchy of levels of processing, applying IPT to olfactory consciousness faces three problems. First, it's unclear whether the olfactory system is hierarchically structured and, moreover, in the same manner as IPT claims for the cases of vision, language, music, and motor control. If there is some manner of hierarchical structure, the question then is whether the same processes are merely functionally realized in a different anatomical manner. Second, even if a retreat to functionally identical levels is successful, it is unclear whether intermediate-level processing does, in fact, occur within the olfactory system. Third, the sensory transduction of odorants—occurring in a combinatorial and nonconceptual manner (see chapter 4)—stands in marked contrast to Jackendoff's information-processing framework, which creates problems for the role of attention, recurrent loops, and top-down conceptual selection.

Currently, it is empirically unclear if the olfactory system is organized anatomically or functionally in an identical hierarchical manner as that of the other modalities. This need not cause concern, however, since it will not get to the heart of IPT's inability to generalize to smell. There might be disagreements regarding the nature of the olfactory system's hierarchical structure, but I have yet to come across anyone denying that it is hierarchically organized. Savic et al. (2000) have shown that there is some hierarchical organization. Yet, it is not at all clear that the organization supports an intermediate level of representation. The most essential claim of IPT is the notion of representational processing and hierarchical stages. Thus, the real test concerns the existence of intermediate-level representations within olfaction processing. The representational levels of processing posited by IPT are at issue with regard to the existence of intermediate-level representations that satisfy the conditions of being perspectival, a bound object,[19] not fully conceptualized or abstracted from the percept.

7.3.5.2 Perspectival representations One of the essential properties of intermediate-level representations is their perspectival format, which is clearly apparent in vision, occurring within a three-dimensional sensory space. Mandik nicely demonstrates how the requirement of having egocentric representations can be adapted for the cases of motor control and thermodynamic representation. However, the body-centric mapping employed by Mandik is not readily apparent in olfaction, especially considering that odorants are usually heterogeneously diffused across a vast area. I might catch whiffs of different smells at different concentration levels of the odorant from point to point in a three-dimensional environment. Yet, it is unclear if my representation of the odorant is relative to a perspective fixed by me or derived from a representation of the environment itself—that is, even if it is allowed on the most permissive account of distal smell perception that we smell odorous objects within smellscapes (chapter 2).

IPT might be adapted to olfaction based upon the role of sniffing and the anatomical structure of the nostrils. It was shown in the introductory chapter how each nostril is physically distinct, which allows us both to track odorants across an environment and to differentiate between aspects of chemical stimuli such as concentration, intensity, and trajectory. Thus, while the notion of olfactory perspective might turn out to be different than the other modalities, intermediate processing theories can nevertheless accommodate this. However, we might need to reconsider the perspectival relation between the object of perception and perceiver relative to each modality proper perceptible, such that slow temporal transduction speed of the olfactory system must be taken into account to allow for a robust account of our experience of smells as particulars within an overlapping array—that is, a smellscape (Young, 2019a, 2019b, 2020).

7.3.5.3 Not fully conceptualized/abstracted from the particular A further underlying tenet of both Prinz's AIR theory and Mandik's AEI theory is that we do not experience the high-level representation. For phenomenological reasons, this claim might be questionable when considering complex odors. When aware of a complex odor, I experience a complex entity, which, as has been argued in chapter 4, is not fully decomposable into its constituent structure (Livermore & Laing, 1998). This aspect of olfactory experience stands in marked contrast to complex visual objects. When viewing a Necker cube, the direction of the lines may be relative to the viewer and seem indeterminate, but one can clearly see the lines and decompose its complex shape into its composite images. When perceiving olfactory mixtures, we can identify, at most, four component odorants within a complex odor, but usually we treat complex odors as their own types. And while Prinz (2012) suggests that this shows it is a tantalizing possibility that smell also has intermediate-level representations, when viewed through the lens of the last chapter that our cognitive smell categories preserve the FNCC format, this might suggest that we are actually conscious of complex smells exactly at the highest level of processing within olfaction.

Our experience of a complex odor is abstracted from its components. Yet, when it comes to simple odors of synthetic molecules, these seem to occur in a manner that is completely unabstracted from their particularity. Moreover, perfume chemists claim to be able to access the smell of an odorant introspectively, while imaginatively combining individual odorants. If their reports are taken at face value, then just thinking about the formula of the perfume gives them an experience of what it will smell like. However, they readily admit that the experience of introspecting a formula is not as robust as actually smelling the chemical complex. Nevertheless, the phenomenon of olfactory imagery further supports this line of thought and shows that olfactory imagery states reactivate the initial nonconceptual and qualitatively conscious sensory and cortical areas in order to elicit the imagined olfactory experience (Young, 2019c). Taken together, these pieces of evidence seem to count against Prinz's intermediate-level representations and Mandik's conceptualist framework, because despite it not being clear at which level olfactory experiences occur and are represented, phenomenologically or otherwise, it doesn't seem to be an intermediate level.

7.3.5.4 Attentional modulation The role of attention within Jackendoff's theory is not essential to his theory, but it deserves further scrutiny. Attentional mechanisms function to attenuate the fine-grained detail of the constituent structure of intermediate representation. However, if the arguments in chapters 4 and 5 are correct, then the nature of olfactory processing is different from that of vision and audition because it occurs in a combinatorial format at the sensory level through higher cognitive levels such that we should not expect attentional modulation to generate more fine-grained representations than those already available at the sensory-perceptual level. While this aspect of Jackendoff's theory is questionable, it can be adapted to a related claim that attention generates stronger links between the representations and long-term memory. However, if there is no explicit constituent structure within the sensory olfactory representations, attentional attenuation seems dubious.

7.3.5.5 Attentional selection What separates Prinz's model from Mandik and Jackendoff is the necessary role of attention in selecting the intermediate-level representation that becomes conscious. The role of attention for smell experience is still an open area of research, but it is arguably the case that one can have a qualitative experience of a smell, even when it is subliminally presented (Young, 2014; Young et al., 2014). Support for this claim can be derived from the phenomenon of blind smell in which subjects are able to detect the presence of an odor, one highly correlated with their confidence ratings, but cannot report on whether they are aware of the odor. Research on the selections of mates and social preferences also demonstrates the role that qualitative smell experience in the absence of occurrent attention can have on our behavior (Young, 2014). Although this evidence generates reasons to doubt AIR's applicability to smell, it does not falsify it, and as W. Wu (2018) has argued, the theory might even be better off by dropping the attentional requirement.

7.3.5.6 Recurrent loops One of the key differences that separates AEI from other intermediate-processing theories is the requirement that there be top-down attentional modulation: we are only conscious of intermediate-level representations when they are modulated by incoming stimuli and top-down processes. There is no doubt that the incoming stimuli-level representations modulate our experiences of smells. Nonetheless, the top-down component is still unclear in humans. D. A. Wilson and Stevenson's (2006) argument that all learning of olfactory objects occurs against a background context of odors, thereby requiring top-down processes for odorant fixation, might be used as evidence that some manner of top-down modulation is required. Thus, Mandik's requirement of recurrent loops should be left open as an empirical matter that seems prima facie reasonable but which requires further research.

I am pessimistic that the reciprocal links between such levels of processing will be similar to Mandik's allocentric–egocentric mechanisms of representation, as these are closely modeled upon stimuli transduction through the thalamus to the sensorimotor sulci of the cortex, which, as we will see in the next sections, generates the inadequacies for the leading neuroscientific theories of consciousness as general theories of consciousness and not just specifically theories of visual consciousness.

7.4 One of These Things Does Not Belong: Hierarchical Approaches Involving the Thalamus and Thalamic Relays

The anatomical structure of the olfactory system presents a problem for current neuroscientific theories of consciousness, which state that a thalamic relay is, or corticothalamic loops are, a necessary condition for consciousness. The olfactory system's unique anatomical architecture and functional connectivity provides reason to doubt Crick's (1984, 1994) theory (Smythies, 1997), Crick and Koch's (1990, 1998, 2005) theory, Koch's (2004) neurobiological theory, global workspace theories (Baars, 1988, 1997, 2002; Dehaene & Changeux, 2003; Dehaene & Naccache, 2001; Dehaene et al., 2006), and the information integration theory of consciousness (Tononi, 2004; Tononi & Edelman, 1998). The anatomical structure of the olfactory system presents a problem for neuroscientific theories of consciousness, which state that a thalamic relay is, or corticothalamic loops are, a necessary condition for consciousness.[20] However, while a thalamic relay may be necessary for consciously analyzing odorants (Plailly et al., 2008), it is not required for consciously detecting or discriminating between odorants (J. L. Price & Slotnick, 1983; J. L. Price et al., 1991b; Sela et al., 2009; Slotnick & Schoonover, 1992; Tham et al., 2009, 2011; Zatorre & Jones-Gotman, 1991). While most other modalities have a sensory thalamic relay between the receptors and cortical processing, the olfactory system has two pathways. There is a primary pathway that projects directly to the orbitofrontal cortex via the piriform cortex, and a second pathway that has an intermediate link from the PC to the mediodorsal nucleus of the thalamus and onto the orbitofrontal cortex (Ongur & Price, 2000). The role the second pathway plays in olfactory processing is reviewed in this section to demonstrate that olfactory consciousness may occur without thalamic mediation.

In rodents, the role of the thalamus in olfactory consciousness is murky, but it is clear that olfactory processing occurs across dual pathways in a similar manner to humans. In rats, there is a similar secondary pathway via the thalamus (J. L. Price & Slotnick, 1983), which is implicated in complex behavioral planning and motor integration. Lesion studies show that it has little or no effect on olfactory discrimination and detection (J. L. Price & Slotnick, 1983; J. L. Price et al., 1991b; Slotnick & Schoonover, 1992). Additional studies of rats show that while lesioning of the thalamic pathway does not affect discrimination or detection or result in anosmia, severing this pathway can produce severe deficits in odor reversal learning (Slotnick & Kaneko, 1981), changes in odor preferences, and male sexual behavior in hamsters (Eichenbaum et al., 1980; Sapolsky & Eichbaum, 1980). Based on animal studies, the thalamus is implicated in behavioral planning and motor integration, and is, to some extent, involved in motivation and attentional mechanisms.

Further research has implicated the role of the MDT in rats as a sensorimotor complex (Courtiol & Wilson, 2014) that subserves foraging behavior and the coordination of learning the sensorimotor contingencies of environmental odors (Courtiol & Wilson, 2016a). Moreover, the MDT in rats shows differential activation for both the type of odorant as well as the sensory pathway of activation that is modality specific (Fredericksen et al., 2019) and might be a necessary component of the working memory system for olfaction. However, since olfactory discrimination and detection are unaffected by thalamic lesions, it would seem that, at least in the case of rodents, the thalamus is not necessary for the realization of olfactory consciousness. Thus, a quick anatomical perusal in animal models demonstrates that the thalamus is not essential for olfactory consciousness in rodents. However, matters are not quite as clear in humans.

Generally, the thalamus is considered partially responsible for attention, memory formation, selective attention, and, to some extent. sensory discrimination (reviewed in Tham et al., 2009), which explains why it is considered by so many as a necessary part of the neural correlates of consciousness. The role of the thalamus in human olfactory consciousness is less than clear due to the sample pool from which evidence is drawn. In animal studies, specific lesions may be generated, but unfortunately evidence for the role of the thalamus in humans must be drawn from a population with brain trauma or general neural deficiencies. As such, the sample size of these studies is quite small, and the lesions are not always clean. In two recorded cases of bilateral dorsomedial infarctions, the patients suffered from abnormalities in perceiving odor character (Asai et al., 2008), which suggests that the MDNT may have some role to play in identifying odors.

Two further research studies, conducted to study the effect of MDNT lesions on olfactory processing, show that while patients with MDNT lesions suffer from deficits in olfactory identification, as demonstrated by their inability to identify an odorant, even on a forced choice task, their ability to detect and judge the intensity of odors is unaffected (Sela et al., 2009). These results indicate that the thalamus may be required for a kind of olfactory awareness that requires the use of one's conceptual repertoire and access to linguistic resources for identifying odors, but not required for the awareness of the presence of an odor and ability to discriminate between odorants. Based on Sela et al.'s (2009) study, the thalamus is not required, with the exception of cross-modal experiences involving conceptual identification for olfaction.

More recently, Tham et al. (2011) have shown that while left-sided MDNT lesions have no effect on odor acuity, hedonics, recognition, naming, and target search, they do have an effect on olfactory discrimination when compared to vision. While these findings differ from the results of Sela et al. (2009) on hedonic judgment and discrimination, the first might be attributed to the sample size and general patient abnormalities, while the latter might be construed as a deficit in contrast to vision. Nonetheless, these deficits are not the result of a general olfactory deficit and, as such, are specific to the role that the thalamus plays in olfactory processing.

The findings of these studies are not completely congruent, but they do indicate that olfactory detection, discrimination, and odor recognition are possible without the thalamus. While the thalamus does not seem to be essential for olfactory consciousness, it does seem to be required for some cross-integration, since each of these studies suggests that the thalamus is a constituent of the olfactory motor system. Lesions of the MDNT do not have drastic effects on olfactory discrimination and detection, but they do affect subjects' abilities to judge flow rates of odorants across their nostrils. This latter finding might be of importance if the sniff is considered as part of the olfactory percept in generating a determination of olfactory quality and odor identification (Kareken et al., 2004; Kepecs et al., 2006; Koritnik et al., 2008; Mainland & Sobel, 2006; Sobel et al., 1999a). Studies on lesions of the ventrolateral thalamus further substantiate the finding that the thalamus is part of the olfactory motor system, since the lack of connection has a negative effect on odor threshold due to decreased motor control and the ability to judge sniff volume (Zobel et al., 2010). Additionally, deep brain stimulation of the cerebellothalamic pathways produces a negative effect on odor threshold and slight effects on discrimination but no effect on odor identification (Kronenbuerger et al., 2010), thereby strengthening the case that the thalamus is part of the olfactory motor system and not necessary for olfactory awareness.

The studies discussed here show that although the thalamus is not required for us to discriminate between odorants or to detect odors, parts of it may play a role in odor identification and motor integration. Thus, the thalamus is not necessary for some olfactory experiences but is implicated—based on its negative impact on olfactory identification—in involving motor integration and conceptual integration. With this more nuanced appraisal of olfactory anatomy and the role of the thalamus in olfactory processing in mind, the next section provides an assessment of the current neurobiological theories of consciousness, all of which claim an essential role of the thalamus. Although the anatomy of the olfactory system and lesion studies of the thalamic relay in olfactory consciousness provide evidence in favor of the traditional view that the thalamus is not required for olfactory consciousness, Plailly et al. (2008) argue that the olfactory system may be similar to the other modalities in requiring thalamic connections. Prima facie, their results vindicate the targeted theories of consciousness based on the conclusion that a thalamic relay is required to analyze smells consciously. However, their study only shows that attending to odors increases the connectivity of the olfactory medial pathway, thus only licensing the conclusion that it is involved with consciously sniffing and attending.

The experimental task of Plailly et al. (2008) was a detection task that required subjects to attend to the presence or absence of an odor in one condition and a tone in a second condition. Subjects were instructed to be attentive and detect the presence or absence of the target. The tone task was used as a baseline to judge the effects of the overall connectivity of the dorsomedial thalamic connections in the olfactory task. Plailly et al. claim that their results of increased connectivity of the dorsomedial thalamic pathway indicate that the thalamus is required when "we consciously analyze smells" (p. 5257).

Given the experimental design, there are multiple problems with this conclusion. The most trifling problem is that their results are overstated, which is evident from the fact that the experiment is a mere detection task from which inference regarding the conscious analysis of smells might be dubious. Properly stated, the results indicate only that thalamic connectivity is increased when attempting to detect the presence of odors. Their conclusion is also unwarranted because the experimental design itself required active sniffing as part of the task. Subjects were instructed to sniff actively for three seconds as cued by a green fixation screen, which allows for an alternative explanation of their data: the increase in the connectivity of the thalamic pathway is probably caused by consciously sniffing, which requires a convergence of motor areas. The evidence suggests that consciously sniffing odors requires thalamic connections but does not show that detecting an odor, while engaging in normal respiratory activity, is not possible without a thalamus or an increase in thalamic connectivity. Rather than substantiated their claim, Plailly et al.'s (2008) findings serve to reaffirm the findings in animal studies that the MDNT is required for complex olfactory motor integration. Additionally, the findings converge with more recent studies on humans, which suggest that the thalamic connection (and gray-matter density) may be increased through smell training, attention, and the use of higher-level cognitive functions as a means of supplementing olfactory sensory sensitivity (Arnold et al., 2020). Moreover, Okamoto et al. (2020) showed that patients suffering from thalamic hyperfusion showed decreases in smell recognition and identification but no change in odor detection acuity. Interpreting Plailly et al.'s result against the background of these more recent experimental findings suggests that the thalamus might play a role in olfactory selective attention, such that there is increased thalamic connectivity when attending to smells. These results are also in line with the findings of lesion studies in animals that show a decrease in performance as the attentional demand of the task increases. Tentatively, this suggests—contrary to previous research on the thalamus's role in attention (McAlonan et al., 2000; Spence et al., 2001)—that olfactory selective attention might be partially mediated by the MDNT.

The medial dorsal thalamus's role in olfactory cognitive processing and consciousness requires further study. Yet, current evidence about the anatomical structure of the olfactory system establishes that thalamic relays and corticothalamic loops are not required for all of our conscious olfactory experiences. The inessential nature of thalamic relays, connections, or loops involved in olfactory consciousness brings into doubt three major groups of neuroscientific theories of consciousness: (1) Crick and Koch's framework for the specificity of the NCC, (2) the GWT of Baars and Dehaene, and (3) Tononi and Edelman's information integration theory. The mere anatomical structure and functional organization of the olfactory system demonstrates that these theories do not provide adequate general accounts of consciousness.

7.4.1 Crick and Koch: Neurobiological Specificity of Neural Correlates of Consciousness

The driving methodological assumption behind Crick and Koch's framework of consciousness is that we should initially assumed that there are specific areas of the brain or specific neural circuits that underlie consciousness, rather than the alternative assumption that consciousness is distributed across the entire brain. The thalamus, with its central location and connections, serves as a good starting point for such specificity. The underlying idea behind the posited involvement of the thalamus is that it acts as a mechanism for the attentional binding of visual information and can create strong reverberatory connections with the cortex.

Crick (1984) claims that the thalamus and the nucleus reticularis are the neural basis of his hypothesized searchlight of consciousness, which is suggested both by the thalamus's topographical maps of the sensory modalities, its cortical loops, and that the reticular nucleus plays a role in unifying our perceptual experiences. Crick thus claims that the thalamus and, in particular, the reticular nucleus are necessary parts of the neural realization of consciousness. While Crick (1994) maintains the instrumental role of the thalamus as the "conductor" that produces consciousness, he is careful to restrict his theory to claims about the NCC of visual awareness. Furthermore, he rejects the intralaminar nuclei and the reticular nucleus of the thalamus as the key to consciousness and replaces them with the lateral geniculate nucleus based on its role in the visual system. Crick admits that his claims regarding thalamic connections do not apply to olfaction (Crick, 1984), but nevertheless assumes that the theory of visual consciousness will generalize across all the modalities. The assurance that these differences need not worry us is given throughout his collaborations with Koch (Crick & Koch, 1990, 1998, 2005) and indeed in Koch (2004). Their general strategy (Crick & Koch, 1990, 1998, 2003, 2005; Koch, 2004) is to generate a framework for understanding consciousness. One of their key assumptions, based on studies of the visual system, is that coalitions of neurons must fire together in circuits to generate enough activation to bind sensory information into a conscious percept. This implicates the thalamus as the seat of attention, since it is necessary for consciously attending to a bound unified perceptual experience. More generally, they claim that the thalamus is a necessary condition of conscious awareness (Crick & Koch, 1998, 2003, 2005), as well as the reticular nucleus (Crick & Koch, 1990), the pulvinar (Crick & Koch, 1990), the LGN (Koch, 2004), and the intralaminar nuclei (Koch, 2004).

Given the specificity of their claimed NCC and the central role of the thalamus therein to bind information attentively, the lack of a thalamic connection within the olfactory system creates real trouble for the claim that this approach generalizes as a theory of consciousness for the other perceptual modalities.

7.4.2 Global Workspace Theories of Consciousness

The anatomy of the olfactory system has the least impact on the GWT of consciousness according to which consciousness is functionally realized by a global workspace system (GWS) that is distributed throughout the brain. Nonetheless, as a neuroscientific theory of consciousness, it cannot remain neutral on the neural realization necessary for global broadcasting. Although the mere lack of thalamic relays within the olfactory system is not decisive proof against the GWT as a plausible neuro-functionalist theory of consciousness, the next section argues that there is no functional equivalent to the thalamus in olfaction, thereby bringing into question whether the GWT can even be functionally generalized to the olfactory system.

Baars's original model built upon the idea that information must be integrated from across the different sensory systems and have access to working memory to become conscious. The integrative property of consciousness is utilized as evidence in favor of there being a global workspace in which information from across the different sensory modalities is combined to form a unified conscious percept. Baars (1988) is explicit in identifying the thalamus as a necessary element in GWS interconnectivity and suggests the extended-reticular thalamic activation system as a possible workspace realization.

Baars (1997; Baars et al., 2003) develops the theory that the thalamus is still implicated as a necessary precondition for waking consciousness in terms of the intralaminar nuclei and the role of the thalamus as a general requirement for consciousness based on a contrastive analysis with other kinds of conscious states. However, these areas are neutral and irrelevant with respect to the issues, for the same reason that the most charitable interpretation of the centrencephalic approach was found to be innocuous and trivial, given its focus on waking consciousness. His current version of the GWT takes the guise of a metaphor of "the theater of consciousness," whose applicability to olfaction is unclear. Nonetheless, Baars's et al. (2003) theory is explicit in endorsing the thalamus as a necessary part in the realization of the global workspace based on its centrality within the brain and its interconnections to the different sensory systems, cortex, working memory, and motor systems. Baars (2013) leaves open the possibility of other functional implementations of the global workspace within the olfactory system and suggests the theory is not necessarily tied to any one form of anatomical realization, as it is open to any system functionally realizing a GWS for consciousness as long as its broadcasting properties fit those of the GWT. However, his suggestions of what these might be do not fit the current understanding of the functional organization and processing of olfaction and are explicitly engaged with above (section 7.4) and in the following section considering functional analogues of the thalamus for olfaction (section 7.4.5). In short, the olfactory system's anatomical structure remains a serious problem for the GWT.

Dehaene's version of the GWT (Dehaene & Changeux, 2003; Dehaene & Naccache, 2001; Dehaene et al., 2006) is indirectly influenced by Crick and Koch's framework via Baars. The neural realization of consciousness is difficult to ascertain in Dehaene's account, since it is offered as a theory not of the NCC themselves, but rather of long-distance neural connections and bidirectional connectivity, and their connection to memory, motor, and language areas, as essential requirements that the neural circuitry responsible for such a global workspace must satisfy (Dehaene & Naccache, 2001). While this is not enough to implicate his GWT in the anatomical crimes of the other theories, Dehaene et al. (2006) suggest a role for the thalamus in this regard, while Dehaene and Changeux (2003) state that pyramidal neurons distributed across cortical and thalamic regions may be responsible for realizing conscious states.

The actual requirements of the GWT model require a role for the thalamus as a necessary condition for consciousness. However, in a more recent review paper, Mashour et al. (2020) note that, in a more recent version of the neural version of the theory (GNWT), perhaps the thalamus is not necessary for consciousness, but it does play an integral role. Thus, the olfactory system's anatomical connectivity does not meet this general constraint, and I would argue that the only option left to them is to retreat to a functionally equivalent model for the olfactory system. Ultimately, if the GWT is merely generating a functionalist account of consciousness in terms of information processing, then it is possible to supplement their claims with a functional analogue, which will be ruled out in section 7.4.4.

7.4.3 Information Integration Theory of Consciousness

The IIT seeks to account for consciousness in terms of the information processing internal to a system. The IIT was proposed by Tononi and Edelman

(1998) and elaborated by Tononi (2004). Its key claim was historically the dynamic core hypothesis, which states that the neural correlates of consciousness are realized by a process of dynamic integration between neural states (for an updated version of their core axioms, see supplementary materials 1 for IIT.3 in Oizumi et al., 2014 and Tononi et al., 2016). Evidence for the IIT derives from Tononi and Edelman's a priori assertion that the two essential underlying properties of consciousness are the integration, or unification, of information (i.e., each conscious experience has some manner of unified content to it) and differentiation (i.e., our conscious experience can rapidly change between drastically different percepts). These properties are used to ascertain the neural processes required to realize informational states capable of generating information integration and differentiation. Tononi and Edelman identify the dynamic core with the recurrent interaction between the anterior and posterior areas of the thalamus and claim that it is required to generate information states that can have the properties of integration and differentiation. While Tononi (2004) does not completely reject this earlier idea, he now only endorses the view that the thalamocortical system is essential for consciousness. As the IIT has developed, the underlying claimed neuroanatomical realization has been watered down, such that IIT.3 only claims that some specific thalamic nuclei might act as enabling conditions for consciousness (Oizumi et al., 2014) or that corticothalamic connections enable consciousness. However, its more worrisome development is championing the back-of-the-brain camp for the realization of consciousness (Boly et al., 2017; Koch et al., 2016) that explicitly ignores olfactory processing or any perceptual sensory qualities, which is especially surprising, since it is meant to explain phenomenal consciousness.

The IIT is the most explicit theory with regard to its endorsement of the thalamus, corticothalamic loops, and back-of-the-brain processes as necessary anatomical hubs for consciousness. Thus, its empirical falsity as a general theory of consciousness is even more apparent, given the unique anatomical structure of the olfactory system. The IIT may work as a theory of the visual modality, but the anatomical structure of the olfactory system simply cannot be accommodated by this approach. Moreover, its a priori axioms tacitly derive from an ocular-centric first-principal bias, as the notion of composition is highly questionable when the format of smell is taken into account, especially as this translates into postulates concerning the mereology of complexes. Since the IIT's anatomical crimes are enough for my purposes, I will leave aside rubbing nonconceptual olfactory qualities into their speculative panpsychic framework.

7.4.4 A Functionally Equivalent Analogue to the Thalamus in Olfaction

It has been argued that the thalamus is not required to be conscious of smells and that the most natural strategy for the theories of consciousness discussed would be to pivot to claim that there is nevertheless a part of the olfactory system that has a functionally equivalent role to that of the thalamus in producing consciousness. More recently, this has been noted as the other systems argument by Doerig et al. (2021), who challenge the assumption that the thalamus is necessary for consciousness and ask us to consider if another functional realizer is possible in other species. Yet, we don't have to go that far, as smell's unique anatomy could easily serve as the other system. In this section, the possibility that other parts of the olfactory system serve as a functional analogue of the thalamus is considered and rejected. The olfactory bulb and olfactory cortex are evaluated as possible role fillers for the functional role attributed to the thalamus as binding information, a common workspace, or integrating information cross-modally.

Since the theories of consciousness are looking for an intermediate-level structure between the cortex and receptor cells of each perceptual modality, one suggestion would be to view the OB as functionally equivalent to the thalamus. Using research on the functional encoding of odorants in the OB, it is shown that the functional organization of the OB is not functionally equivalent to the role assigned to the thalamus within these theories.

Using the intermediate stage of processing approach, Kay and Sherman (2006) argue that the OB is functionally equivalent to the thalamus—that is, it plays the same role in the olfactory system as the thalamus in the visual system—on the basis of three claims.

The first claim is that both the OB and the LGN are anatomically situated at an intermediate stage of processing between the receptor cells and the cortex. However, this observation does not support the claim that the OB and LGN are functionally equivalent and only supports the much weaker claim that if vision has three stages of processing projecting to the cortex, so might the olfactory system (depending upon whether the OB is considered a receptor site similar to the ganglion cells of the retina or the LGN of the visual pathway).

The second claim is that both the OB and LGN serve as a bottleneck within the informational stream that reaches the cortex. The popular metaphor of an information bottleneck is best unpacked as the claim that both structures focus incoming stimuli by decreasing the amount of information projected from the receptor sites to the cortical areas. However, this falsely assumes that the functions of glomeruli and mitral cells within the OB are to act only as relays, and thus grossly underestimates their actual functional role.

Lastly, Kay and Sherman (2006) use the structural similarity of mitral and tufted cells of the sensory input circuitry within both the OB and LGN as a means of comparison. While in general understanding structural organization facilitates a better understanding of function, it is essential in this case to take into account the actual workings of the OB at a more detailed level.

Given these problems, it is quite reassuring that Kay and Sherman admit ignorance regarding the functional role of mitral cells within the human OB. But we might be able to elucidate this based on some dated research from the zebrafish animal model of olfaction. Friedrich and Laurent (2001) use the zebrafish as an animal model for OB function in humans. Based on the convergence over time of olfactory receptor cells firing rates and, in particular, the convergence of firing patterns within the odor-coding assemblies of mitral cells in the OB of zebrafish, they suggest that the OB encodes odorants in a combinatorial manner, such that the representation of a stimuli is holistically encoded in the firing patterns of the glomeruli and mitral cells across the entire OB itself (rather than each aspect of the stimuli being discretely encoded within it). This nicely highlights a key difference between the functional organization of the olfactory system and all other perceptual systems in terms of the variable of time, as well as its divergent compositional format that it unlike the isomorphic encoding strategy seen within the visual system as it projects from the retina via the thalamic relay in the LGN to V1. Taken together with the problems raised for Kay and Sherman's three claims, it looks unlikely that the function of the OB should be equated with that of the LGN of the thalamus.

Another reason to reject the claim that the OB is functionally equivalent to the LGN is that Kay and Sherman only compare the LGN to the OB. Consequently, their results are methodologically sound only if all areas within the thalamus do in fact function in the same manner as the LGN—something that needs to be demonstrated. Even if, as Kay and Sherman claim, the OB is, in fact, functionally equivalent to the LGN, this will only assist the neurobiological approaches of Crick (1994) and (perhaps) Koch (2004): it would be of no help for the GWT, the IIT, or any other theory according to which the general functional role of the thalamus is to bind information cross-modally.

Ascending the hierarchy of the olfactory system, the natural place to look next for a claim of functional equivalence is the olfactory cortex. Murakami et al. (2005) have shown that the state-dependent gating mechanism in rats, which occurs at the thalamus for all other sensory systems, can be seen to occur at the anterior piriform cortex and olfactory tubercle of the OC. This demonstrates that sensory gating occurs within olfactory processes, and that the sensory gating with the OC is in synchrony with the activity of the gating mechanisms of the other modalities located in the thalamocortical system. While this might indicate that the APC and OT are the functional equivalents of the thalamus in olfaction, this would overstate the results of Murakami et al. Apart from the fact that these results are only from animal models, sensory gating at best shows that the olfactory system employs the same mechanisms for information processing of incoming stimuli and not the full range of properties attributed to thalamic processing by the theories surveyed above. Despite the results showing that olfaction must at times work in concert with the other modalities, they do not yield the full equivalence of function to the thalamus posited by the neurobiological theories of consciousnesses covered in this chapter.

While only two possible candidates for a claim of functional equivalence have been considered, to the author's knowledge, there are no other extant theories that claim that there is a functional equivalence between the thalamus and parts of the olfactory system. Attempting to accommodate olfactory consciousness within the theories of consciousness considered in this chapter is an ill-advised research strategy: olfaction works differently from the other perceptual modalities, and the functional organization of the olfactory system and its encoding mechanisms from the receptor sites to the cortex are unique in many ways. The burden of proof thus lies with those theorists who make a claim of functional equivalence either to show that all the areas of the thalamus function in the same way as the LGN or to posit an alternative structure of the olfactory system that could fulfil the claimed functional equivalence. The likelihood of the first option is extremely dubious, and the possibility of the second option requires neuroscientists to engage with the workings of smell if they wish for their theories not to be false or inadequate as general theories of human consciousness.

7.5 Conclusion

As the most elemental sensory modality, olfaction holds the key to a fundamental understanding of consciousness and the qualitative character of experience. Olfaction provides novel explanations of the qualitative character of consciousness based on the olfactory object being the chemical structure of molecular compounds, that olfactory sensory, perceptual, and cognitive states are nonconceptually formatted, that contemporary neurobiological theories of consciousness are inadequate, and that the qualitative character of consciousness arises from sensory states, which are necessary for awareness and which occur in a nonconceptual format.

The study of consciousness as an empirically viable research field is no longer nascent and needs to move beyond its toddler developmental stage of ocular fixation. If scientists are unwilling to wake up and smell olfactory consciousness, at least they could note this bias explicitly when stating general claims about how the brain generates consciousness of whatever kind. What has been shown is that swaths of empirically viable theories are still inadequate because they suffer from an attentional neglect of smell. My own broad sweeping claims themselves must be modulated, since I have only surveyed a range of the leading theories and none of the explicitly philosophical approaches. Nevertheless, as consciousness studies continue to progress, I hope the field realizes that studying smell will yield fruitful results—it certainly has within olfactory philosophy over the past decade, as can be seen in hindsight when the conclusions of all the previous chapters are tallied.

The olfactory system's unique nature makes it ideally situated to make novel predictions about qualitative conscious experiences. For instance, the last chapter showed how olfactory consciousness is special because qualitative consciousness occurs in the absence of awareness. Yet, all states of olfactory awareness have a qualitative character. But some even bigger conclusions are possible when the findings of the proceeding chapters are combined. If I were to oversell my *Stinking* findings, I might claim that studying smells warrants the conclusion that olfactory consciousness derives from our interactions with the basic elements of reality whose qualitative character derives from the nature of chemical reality, and because one's olfactory experiences are nonconceptually formatted, the nature of olfactory experiences cannot be predicted based on concepts. Furthermore, whenever one introspects, thinks about, remembers, or is directly aware of an olfactory experience, these states contain an olfactory quality, which necessarily requires the reactivation of the original sensory areas. However, given the magnitude of such claims, it is best not to oversell olfaction. Hopefully, you will be able to sniff out these conclusions and their implications for the metaphysics of mind or accept a promissory note until I can write a book on Anosmic Annie.

Smell you later!

Notes

Acknowledgments

1. I would also like to credit my picnic table office spaces over the years for inspiring some of my writing: Palmachim Beach, Tahoe Vistas, Carnelian Bay, Sand Harbor, Linda Mar, Stinson Beach, and Sand Dollar Beach, Plaskett Creek, as well as the Big Sur Bakery, whose blueberry strudel served as the writing incentive for multiple portions of this book.

Chapter 1

1. Despite making broad statements about smell, this book is primarily concerned with and focused upon human olfaction. The limited scope is partially because we are most familiar with our own sense of smell, and thereby this forms an easier starting point to theorize about the nature of perceptual experiences, capacities for cognizing, and communication about smells, as well as our consciousness. Animal models of olfaction will be set aside and brought in only when directly relevant, which is not to discount their relevance or importance. Each animal model of olfaction is a specialization in its own right, and most are beyond the remit of my expertise. Additionally, covering them in any detail would make the book cumbersome in length.

2. For good state-of-the-art reviews, see S. C. Roberts et al. (2020) and Schaal et al. (2020).

3. For further details, see section 7.4.3.

4. For a recent study demonstrating the role of mucosa enzymatic clearance of odorants that are further implicated in behavior of neonates, see Robert-Hazotte et al. (2019b).

5. Recent research has shown that in addition to changes in environment smellscapes modulating ORN neurogenesis, the enriched odor environment as well as

high respiratory rates for the novel odorants might be drivers of neurogenesis of the mitral and tufted cells within the olfactory bulb (Kamimura et al., 2022).

6. For a good recent introduction to the central olfactory structures with clinical implications, see Cleland and Linster (2019), and for a recent study focusing on functional connectivity of olfactory cortical processing centers, see Arnold et al. (2020).

7. Limbic connections are direct and unmediated. Some argue on this basis that olfaction has a stronger effect on memory encoding than the other modalities. However, it is unclear whether olfactory stimuli create stronger emotional responses to their presence or cause stronger memory encodings, or whether remembering olfactory experiences is more vivid. Since nothing argued for within the book turns on this issue, it is set aside as a topic of interest along with the emotional mediation of smell perception for further research.

8. Hopefully, I will get to this last topic soon, given my outstanding title "Fragrant Violations" that needs a paper to accompany it.

Chapter 2

1. The chapter will set aside general metaphysical issues such as the ontological status of smells for such a treatment; see Cavedon-Taylor (2018).

2. A driving assumption of the current framework is that these questions are nested such that an answer to one has implications for the others.

3. The comprehensiveness of this account is limited to orthonasal smell. Elsewhere, it is argued (Young, 2023) that retronasal olfaction should not be considered part of the modality of smell. By limiting smell in this fashion, some might argue that the framework is already limited in scope and not very comprehensive, but for good theoretical reasons developed in Young (2023), it is in keeping with our pre-theoretic conception of the senses and well supported if we start with the assumption that the primary determinate of a smell's identity is its olfactory quality.

4. Note different ordering and phrasing of questions—previous versions of the question address a different issue that will be handled in chapter 4.

5. Elsewhere, Young (2019a) addressed the strengths and weaknesses of the full range of philosophical theories within the literature on smell in answering these three questions. Here, the focus will be just on stress testing my own theory in handling criticisms and suggesting further issues that still need to be addressed.

6. For the purposes of the chapter, I am adopting the chemoscientific linguistic convention that odor is used to refer to the olfactory quality of a smell often

characterized in terms of our subjective experience of the chemical stimulus, while odorant is employed to refer to the chemical stimulus.

7. Throughout the book, "odor" will be used in a limited sense to refer to the olfactory quality of odorants, while "smell" will be used in a more expansive sense to include other aspects of olfactory perceptible object beyond its olfactory quality.

8. More nuanced issues will arise if detection is employed as the only means of odor identification, since detection thresholds can be modulated based upon extraneous background conditions, yield specific individual differences between men and women (as well as fluctuations within menstrual cycle), and can be modulated based on external perceptual modalities such as vision. For instance, the presentation of disgusting images lowers the detection threshold for subsequently presented pleasant and mildly negative odors (Chan et al., 2019). Additionally, in the case of super smellers, whose capacity for odorant detection is exceptional, the heightened ability seems to be driven not by an enhanced receptivity at the sensory level, but rather by increased gray matter in secondary olfactory processing centers in the brain (Wabnegger et al., 2018).

9. Recent research suggests that subjective differences in odor perception must also be taken into account in theorizing about odor identity and categorization, given individual variance in subjective perception and reported phenomenology is rife and has been ignored in past research studies (Mantel et al., 2021).

10. In keeping with contemporary chemoscientific practices, olfactory perception in humans is specified and limited to the olfactory sensory system and the proper sensible of smell, where the latter is the issue at hand. The olfactory sense organ starts from the olfactory epithelium, including the mucus layer, extending through cortical processing within the piriform cortex and olfactory cortex (for a short introduction to the olfactory system, see Bensafi et al., 2004, and Young, 2011). Sensorimotor sensations of breathing, sniffing, and the tactile stimulation of the nasal cavity will be excluded from consideration as part of the olfactory system, as will trigeminal stimulation. Although trigeminal stimulation influences the phenomenal content of reported smell experience, the trigeminal system has its own sensory qualities (Filiou et al., 2015). Thus, instances of trigeminal stimulation influencing olfactory quality should be considered as cross-modal effects.

11. Since the olfactory system is ontogenetically ancient, a great deal can be inferred about human olfactory experiences from animal models. Aside from the fact that it is a generally accepted practice throughout the sciences, Aristotle (DA II 7 $419^{a}33$–$^{b}1$, DAII 9, De Sensu 5) also supports the methodology of thinking of olfaction as lying on a continuum with other organisms. Animal models do not provide any further evidence for the claim that olfactory objects are the structures of molecular compounds; they only reinforce it. So, for the sake of brevity, evidence from animal models has been omitted.

12. The most exhaustive list of the smells of enantiomers is maintained by John Leffingwell. According to calculations of Leffingwell's listed in Turin (2006), 64 percent of enantiomers smell the same, 17 percent smell different, while the smell of the remaining 19 percent is currently unknown.

13. Odor concentration invariance, whereby we recognize an odor as having the same olfactory quality across presentations of varying concentrations, is an acquired perceptual capacity, which depends upon learned odor categorization. Concentration invariance extends beyond the perceptible properties presented by the external object of olfactory perception, including its property of olfactory quality. Although further research on olfactory perceptual constancies of concentration invariance in humans is needed, it is most likely determined in accordance with the ratio of the chemical compounds within a given odor mixture, since the compositional ratio of components should stay constant, despite a shift in concentration levels (Uchida & Mainen, 2007). Given that we have to learn to identify an odor from across multiple presentations of shifting concentration that generate discriminable differences between tokens of the same odorants at different levels of concentration suggest that both top-down processing as well as the odorant plume as a superordinate perceptual object must be included within any account of the determinates of what makes an odorant have a particular smell.

14. While we are beginning to gain a better understanding of the nature of the sensory transduction site and how odorants enervate the receptor, we still don't know how many different types of odorants can bind with the receptors on any given neuron at any one time or how the promiscuity is controlled for upstream in encoding odorant identity at the olfactory bulb. Moreover, the receptive range of the same type of receptor is not conserved across species such that there is variance in receptivity between species. For example, *Drosophila* ORNs show differentiation for similar odorants, which increases their ability to generalize across instances, while locusts separate ORNs for similar odorants yet misclassify variants of the same odor (Rajagopalan & Assisi, 2020).

15. For arguments and a model that SOR frameworks incorrectly assume a linear and isomorphic match of properties of the stimulus to perceived odor qualities, see Y. Zhou et al. (2018), who argue that hyperbolic geometry provides a better model for mapping olfactory quality space.

16. Skrzypulec (2018) offers a detailed analysis of mereology, comparing and contrasting olfaction and vision. Based on his analysis, olfaction has mereology in terms of part–whole relations of complex olfactorily mixtures. Yet, it is it not classical in its structure and differs from the mereology of vision.

17. Budek and Farkas (2014) offer an alternative strategy to argue for the olfactory objecthood based on the phenomenal presence criterion: "the object of perceptual experience is what seems to be present when having the experience" (p. 11).

However, I set their treatment aside, as its phenomenological starting point using conscious awareness seems to get things wrong from the outset, given that most olfactory experiences occur in the absence of conscious awareness (see chapter 6) and there are serious worries as to the reliability of introspective reports about what we think we experience when theorizing about smells (Young, 2019b).

18. While the very idea of odor objects has been criticized as an unnecessary and unmotivated posit (Barwich, 2019; Keller, 2017), these criticisms have been addressed in Young (2023).

19. For a philosophical analysis of how our experience of odor hedonic and valence, particularly with regard to disgust, provides an alternative account of the intentionality of olfactory experience as imperative intentional content, see Martinez (2015).

20. Since the relation between odor concentration and quality has been handled in section 2.2.1, it is left aside here in favor of focusing upon supporting the claim that a smell's intensity can be treated as a discrete perceptual dimension of our experience of smells as complex perceptual objects.

21. While some philosophers claim that smells just appear within our noses and are sensory ephemera Richardson (2013b) offers a strong argument against such views and that in fact smell should be consider an extrooceptive sense. Moreover, a recent study by Y. Wu et al. (2020) showed that we can use stereo-olfaction in determining self-movement relative to the concentration ratio of odorants between nostrils. What is of further interest about this study is that this capacity occurs in the absence of conscious awareness and verbal reports of the subjects to the fact that they could detect the ratio of concentration is olfactory navigation, which is in keeping with what has been argued in section 2.2.1 that verbal reports and conscious awareness should not be consider the gold standard for determining what smells are.

22. Detailed evidence and arguments for these claims can be found in Young et al. (2020) and Young (2020). For more recent evidence, see Raithel and Gottfried (2021), whose review of olfactory navigation strategies corroborates that humans can use smell for navigational purposes and have olfactory spatial mapping capacities.

23. In what follows, I set aside Comanski's (2020) naive-topology framework that can account for the spatiality of smells within a synchronic time frame. On his account, we can have spatial representations that do not include the objects distance or direction. Additionally, I set aside Aasen's (2018) framework that allows for both synchronic and diachronic experiences of odors as having distal properties.

24. For more elaborate treatments and criticisms of his overly strict construal of the accuracy conditions of olfactory perception as derived merely based on the receptivity of the olfactory system, see Young (2019a, 2023), as well as his arguments deriving their strength from analogy from object perception in vision (Young, 2019b).

25. The many-property problem is proposed as an explanatory challenged in generating an account of how we can perceive the same types of properties instantiated in different arrangements across a variety of perceptual arrays (Jackson, 1977; A. D. Smith, 2002; Tye, 1989).

26. It is acknowledged, however, that further research is required in studying how the odorant composition of an odor plume modulates our perception of the resultant smell's olfactory quality.

27. For example, trigeminal stimulation from capsaicin that co-occurs with gustatory perception is often thought to generate flavorful perceptual qualities of pungency.

Chapter 3

1. For further arguments and evidence that orthonasal olfaction should be considered our sense of smell, while retronasal olfaction, despite sharing the same stimuli and receptors, should be considered a separate sensory system that is not part of what we consider our sense of smell, see Young (2023).

2. Attributed to S. Firestein by Shepherd (2012) and introduced to me by Barry Smith.

3. Hummel et al. (2006) provide a fantastic review of the differences in perception between orthonasal and retronasal perception, including context of presentation, airflow, trigeminal stimulation and ORNs.

4. The rubric of contemporary chemoscience regarding gustation expands the typology of basic tastes to include the sensory quality of umami that is typically found throughout Japanese cooking, as well as some variety of metallic flavors (for a good discussion, see Shepherd, 2012).

5. B. C. Smith (2015) does give an example of menthol flavor, which might allow us to tease apart the different olfactory, trigeminal, and taste qualities within a given presentation, but admits that this is an outlier.

6. Similar strategies have been employed to establish the representation status of smell (Young, 2016) and the objective nature of olfactory perception (Millar, 2017).

7. However, it might be noted that outside of extreme laboratory conditions, we do not experience the sensory qualities of food as occurring within the nose (Heilmann & Hummel, 2004), but rather we refer the experience as occurring within the mouth (oral referral).

8. Oversimplifying, pica is an eating disorder whereby people consume items that are not considered foodstuff and which contain little to no nutritional value.

9. While the author wishes they had come up with this wonderful example of remembered tastes of nonconsumable flavorful items, it must be attributed to Maria

Larsson, who uses it to demonstrate the encoding and retrieval strength of odor and taste memories.

10. Deroy (2009) is similarly not sympathetic to the subjective account. Her dispositional account is generated contrastively by arguing against the idea that the flavor profile of a wine might be reducible to its chemical composition. Deroy argues that we can separate the chemical properties of a wine's flavor profiles from its chemical composition because these can be known independently of each other. Presumably, the argument should be construed as depending on an epistemic claim, whereby our knowledge of the chemical properties is not sufficient for generating a complete description of our experiential access to the flavor properties. To generate this conclusion, Deroy employs a distinction between objective and evaluative properties of a perceptual object. The objective chemical properties of the wine are deployed as a means of accuracy testing claims regarding evaluative claims of the perceptible properties of a wine. The chemical profiles of a given wine and its development can be used to assess individual's claims reliably about the flavor profile of a wine. However, the flavor profile cannot be reduced to our knowledge of the wine's chemical components. Understood in this manner, the argument concerns what epistemic access is warranted from our knowledge of a wine's chemical composition.

Her strongest evidence for this claim is that two wines, which share some of the same chemical attributes, might not have the same flavor. Moreover, the argument piggybacks upon the assumption of the multiple realization of the olfactory qualities, given similarities and differences between chemical compounds. However, simply knowing something about a few components (as per her own examples) is not sufficient for generating a full account of all the flavorful properties of a wine. We should anticipate that different chemical properties of components within any given bottle of wine might shift the flavor dimensions of the wine. Even if there is an overlap of chemical components, given that flavor is a multisensory perceptual modality that generates a unified synthetic percept, the conclusion that we cannot know the full flavor profile from our knowledge of the chemical composition of a few components does not follow. Despite an overlap of chemical components, the perceptual experience can shift if the background mixture and testing conditions are not held constant across presentations (even in instances that we know all of the chemical components; B. C. Smith, 2008). Since the same intervening variable that influences flavor perception can be accounted for by the more explicit objectivist approach, I think it is best to stick with B. C. Smith's (2009) claim that fully perceiving the robust flavor of a wine requires a level of skillful expertise in forcing the wine to reveal its full objective character.

11. In a number of places, B. C. Smith (2008, 2009) argues that flavor perception is a skillful action requiring us to probe the perceptible object diachronically. The act of tasting forms a performative role in discovering and unleashing the full range of flavor properties. Considering tasting as skillful action fits with sensory research showing converging cortical processing across a range of areas that underly

gustation, retronasal olfaction, somatosensation, tactile and thermal encoding, as well as cells with convergent receptivity in the orbitofrontal cortex (Stevenson, 2009; Blankenship et al., 2018; Goldberg et al., 2018). Moreover, claiming that tasting is a skillful action justifies his further claim of different access points to the wine's flavor between novices and wine experts. Interestingly, the convergent sensory cells in the OFC might also explain why even wine experts have difficulties teasing apart the gustatory, somatosensory, tactile, and olfactory qualities of a wine (Stevenson, 2009a).

12. While research on the representational nature of complex olfactory mixtures derives from orthonasal smell, there are good reasons to think that it is generalizable to retronasal olfaction as well (Shepherd, 2012; Stevenson, 2009a).

13. The existing literature on expertise in chemosensory perception suggests that the greatest indicator of improved performance on detection, discrimination, and identification task is the sheer number of previous experiences (reviewed in Young, 2019a, and Young et al., 2014).

14. Macpherson (2011b) notes two possible approaches to flavor as cross-modal yet fully integrated informational states that feature sensory integrating within the perceptual state or purely unimodal flavor experiences. The theory on offer sides with the latter option, as not only do the sensory qualities presented within flavor experiences differ from their perceptual qualities when perceived independently, but moreover, perceptual states of flavor have a synthetic representational format that does not allow access to the individual components of gustatory or retronasal sensory states.

Chapter 4

1. Burge's (2010) argument that olfaction cannot even amount to a perceptual system based on Salmon's inability to generate perceptual constancies will not be considered because Carvalho (2014) provides a forceful and convincing reply to Burge.

2. For an excellent analysis and overview of Reid's treatment of olfaction, see Quilty-Dunn (2013).

3. Further evidence that olfactory experiences are representational can be derived from those that posit olfactory objecthood such that it satisfies figure–ground segregation (Young, 2016) or principles of gestalt psychology (Millar, 2017), together with the detailed analysis showing that depending upon the theory of olfactory objecthood, endorsed smell perception can occur with amodal completion (Young & Nanay, 2021).

4. See Young (2014) for fuller discussion of the use of secondary measures for determining the sameness of olfactory quality across perceptual and conscious states.

5. It should be noted that the discussion of the representational status of olfactory experiences is meant to be as neutral as possible. I am neither taking a stance on the essential condition of representational status nor stalking a position within the debate regarding what can be perceived (objects and/or contents of perceptions). Rather, I am merely surveying a range of options and showing how olfaction can satisfy them. Perhaps in the later example of perfumes, these examples could be interpreted not as perceptual but rather as post-perceptual cognitively mediated propositional states. Nonetheless, that would still locate them as having representational format, and my examples concern experiences and not explicitly just perceptual states. What the section aims to do is show that olfactory experiences are representational and nothing further about the content of perception (for a good overview of the range of options, see Siegel, 2021). If inflated views of perception such that we can perceive high-level properties, categories, propositions, or forms is your thing, then olfaction can most likely do that. And if your intuitions err on the more sensory side, then these examples of perfume might just be post-perceptual propositional states. Nonetheless, the proceeding examples in sections 4.1.1 and 4.1.2 should provide ample evidence of olfaction's representational status.

6. Based on the chemical nature of the target domain of olfactory stimuli, it might be possible to argue that the target domain essentially requires a novel means of representation. The nature of chemical interactions between odorants is of import in both encoding and representing olfactory experiences. Thus, encoding and representing chemical mixtures will prove more difficult because, at the level of chemical analysis, these stimuli have synergistic properties and behave in a fashion that is unlike the mere summation of their constituents (Earley, 2005). However, this argument will be set aside and not developed in what follows.

7. An account of the debate regarding the role of the piriform cortex in olfactory perception and its implications for philosophical theories of the objects of olfactory perception can be found in Barwich (2016).

8. The current section's argument and evidence derives from mappings and encodings of stimuli as inputted to the PC, which suggest the coarse-grained distributed functionally compositional representations of complex odorants. However, the pathways between the PC and the orbitofrontal cortex suggest that there is a degree of organization and spatial topography to odorant encoding within the PC in connection with the OFC (Chen et al., 2014). Since the issue at hand concerns the pathways relevant to odor language encoding, these pathways between the PC and OFC will be left aside. Olofsson and Gottfried's research (see section 4.2.2.3) indicates that the projection to the ATL is from the PC. Thus, the role of the OFC in this regard is not clearly of relevance.

9. These claims are relativized to English speakers. For an in-depth discussion of the cultural and linguistic mediation of our ability to categorize and identify smells, see the next chapter.

10. For a discussion of different methods of increasing olfactory encoding that yield increased descriptive accuracy in olfaction but not necessarily identification, see section 4.5.

11. In addition to the difference in accuracy at identifying odorants, Savic and Berglund (2004) show that there are separate cortical areas responsible for processing familiar and unfamiliar odors.

12. The paper proceeds under the general assumption that our olfactory capacities for discrimination and identification are linked. Thus, by comparison, there is a discrepancy between them. However, one could challenge this assumption and claim there is no puzzling discrepancy. Rather, we require two independent explanations: (1) what it is that facilitates olfaction's superior discriminative ability for perceptual qualities when compared to other modalities, and (2) why we are bad at identifying smells by name. This alternative will not be entertained within the paper, as a more parsimonious and comprehensive single explanation can be provided, assuming that capacities are linked. Both capacities and the discrepancy between them can be explained by noting the compositional encoding strategy at the sensory level of olfactory processing, which generates our efficient discriminative abilities. Yet, the representational format of some higher levels of olfactory processing, such as olfactory perception and cognition, is incompatible with the format of semantic conceptual systems.

13. Gerkin and Castro (2015) have disputed the methods used to generate this number, and they claim that the original study demonstrates that we can detect at most 1.72 trillion odors. Even if their criticisms are accepted and the original model is corrected, the estimated number of odors that humans can detect would still vastly exceed our ability to identity them, which would still yield the puzzling discrepancy. A detailed discussion of the vast dimensionality of odor quality space using judgments of just noticeable differences can be found in Young et al. (2014) and Keller (2017).

14. Keller (2017) accepts the line of argument that insufficient connectivity to language centers sufficiently explains the discrepancy between our naming and discriminative abilities, but he claims this is in keeping with the overall function of olfactory perception. His argument is that olfaction is selectively designed for the guidance of action and behavior. Thus, it has greater connectivity with cortical areas responsible for these functions. Nonetheless, his account does not offer an explanation of why the lack of connectivity on its own is sufficient for explaining the discrepancy. Moreover, it is unclear how his account can handle the cultural differences in olfactory naming abilities (see chapter 5) without a risk of confabulating the selected function of olfaction with its current function.

15. The alternative explanation offered in this paper derives from both the olfactory system's interconnection with language centers and the format of representations

employed in processing at these hubs. Explaining the puzzling discrepancy in a comprehensive manner requires considering the olfactory pathways, cortical connections, and representational formats at each stage of processing. In an ideal hypothetical system, everything and anything can be connected and processed along a host of different pathways and formats. However, using olfaction as the model system in this instance constrains the realm of possibilities. For instance, the constituents of compositional odor representations cannot simply be decomposed and inputted into processing by the conceptual cortical systems because of a lack of direct and unmediated pathway between these processing areas. In fact, even when complex mixtures yield elemental mixtures (see section 4.4.2), this is subserved by independent olfactory pathways in the amygdala.

Chapter 5

1. While this might be thought just to be implementing an exemplar or statistical regularity theory of concepts, careful analysis bears out that groupings are often arrived at based on holistic judgments about the combinatorial representation of the mereological composition of complex of the stimuli.

2. Despite their interesting findings, it might be worth noting the self-reported flaw in the design of how exact the primary odor tags are, as these were not initially validated.

Chapter 6

1. Perhaps it could be argued that my taxonomic kinds could be better captured by keeping Block's distinction and adding the further distinction between access and accessible (Wu, 2018), but I find that there is a general tractability issue with access that is further obfuscated with the additional notion of accessible yet not accessed. Moreover, this further elaboration has always left me puzzled as to how we can clearly delineate states that are accessible but where the subject is unaware from plain good old-fashioned experimentally reputable notions of unconscious cognitive processing (see Goldstein & Young, 2022).

2. High-level cognitive states are left aside in what follows, since this gets more to the methodological issues of ascertaining the nature of olfactory categories.

3. I am grateful to a reviewer who pointed out that, given the vast dimensionality of smell's sensory qualities, there is currently no consensus on how to create a quality space for smell successfully. As a tongue-in-cheek reply, I must note that as one of the three philosophers who has published on olfactory quality space, I certainly agree that there is no consensus about how to set up the space. There never will be a consensus on just about anything in philosophy, even when the three philosophers publishing on this topic are friends and co-authors. But I am unaware of anything

published that sheds doubt on our proposal(s) being theoretically and experimentally impossible. On a more serious note, there are disagreements between us on whether the JND measurement should be a pairwise or a triangular comparison task and which sensory properties should be selected for as a means of discrimination. If the participant is asked to discriminate between stimuli based on olfactory quality, this will generate a narrow set of dimensions and be more fitting with MST's determination of odor identity in terms of quality (see chapter 2). In contrast, if the JNDs are made in terms of the stimuli's smell than given my broader notion of these as encompassing odor quality, valence/hedonics, intensity, and concentration (chapter 2), then the dimensionality of the quality space for smell will be more robust. Both approaches can be encompassed by my framework within this chapter, and there might be interesting theoretical and experimental difference that could be explored in the future. Yet, any possible skepticism about the difficulty of setting up the dimensionality does not detract from the argument and evidence that qualitative consciousness dissociates from conscious awareness, although the opposite is not the case.

4. The nature and debate regarding human pheromones are irrelevant to all claims regarding the olfactory mediation of human mate selection within this chapter, as all the evidence offered derives from olfactory detection from the olfactory epithelium through higher levels of olfactory processing. The phenomenon under discussion in this section does not conform to the definitional nature of pheromones and is not mediated by the vomeronasal system, as is the case in other mammals (for a more in-depth treatment of pheromones, consult Doty, 2010).

5. These studies of anosmia do not specify the nature of the anatomical damage, since their focus is upon the resultant olfactory deficit. Thus, to test fully the claim that qualitative consciousness occurs at the sensory level in olfaction in the absence of awareness, further research is needed on anosmia resulting from a severed olfactory tract.

Chapter 7

1. To precisify this point, Storm et al. (2017) review only visual evidence for their claim that consciousness is in the back of the brain yet extrapolate universal claims about the cortical localization of consciousness. Moreover, their Perturbational Complexity Index (PCI) measure is highly dubious if not limited to visual consciousness.

2. My own conception of awareness from chapter 6 is neutral enough to capture this general notion, but for the purposes of what follows, each theory will be judged on its own terms, thus setting aside any fine-grained ax-grinding.

3. The relationship between this type of consciousness and the previous ones is a contentious debate that is handled in the previous chapter.

4. The simplest understanding of qualitative consciousness is that, from our own perspective, we can notice that experiences are different. There's something that it is like to have a cup of hot chocolate on a cold winter's day inside the MET café overlooking Central Park or hearing Mahler's Symphony No. 1 being played by the New York Philharmonic on the Great Lawn on a hot summer evening at dusk—these experiences differ in some aspect of the quality of the experience.

5. It might be worried that his theory applies to access consciousness in addition to waking consciousness for the reason that any mental state that guides behavior is traditionally thought to have a content that is available for the guidance of behavior and/or inferential process. However, all he is minimally claiming is that some states occur without awareness yet can play a functional role in guiding behavior. His claim regarding functional role and behavior amounts to nothing more than the common claim that sometimes unconscious processes guide our actions.

6. Merker's centrencephalic proposal might be criticized because Penfield and Jasper's research concerned epileptic patients who were awake and reporting on their conscious experience during surgery. However, his theory concerns the mere preconditions of having an experience, irrespective of whether one is aware of undergoing any experience. Thus, it might be argued that he is conflating waking consciousness and awareness. Nonetheless, if the cortex is not required for even the later kind consciousness, this certainly adds further credence to the centrencephalic proposal.

7. The assumption that these children undergo qualitative or phenomenal conscious states might seem contentious to some but is perfectly in keeping with Block's conception of p-consciousness, Rosenthal's thin phenomenality, and olfactory qualitative consciousness (see chapter 6).

8. Sections 1.3.4–1.3.6.

9. For a full discussion of, and evidence for, this point, see section 7.4.

10. For an introductory coverage of the dissociation between consciousness and attention, see Wu (2018).

11. The definition of access and phenomenal consciousness is carefully explicated in chapter 6.

12. For a more detailed discussion, see section 7.4.1.

13. The rest of the section focuses upon the applicability of Lamme's notion of awareness to olfaction. Lamme's approach to visual p-consciousness might apply to olfaction, so long as it can be ascertained whether phenomenally conscious olfactory states occur without attention, which will be partially touched upon in the next chapter.

14. As noted in the apologetic introduction, the book will not offer an in-depth coverage of olfactory memory and so leaves this issue here for future research.

15. A detailed discussion of concentration as a property of the olfactory object and its ability to change the olfactory percept through increased neural activation can be found in chapter 2.

16. Jackendoff's argument against this view is simply that it does not allow concepts to be modality specific. For the full argument, see Jackendoff (1987, p. 286).

17. For his argument regarding the form of linguistic awareness as derived from the level of phonological awareness and the tip-of-the-tongue phenomena, see Jackendoff (1987, pp. 287–292). For arguments regarding the form of musical awareness involving notes, not pitch or cochlear encodings or concepts as derived from contemporary classical music involving language/verbal structures, see Jackendoff (1987, pp. 292–293). For his argument regarding the form of awareness in visual perception being the 2½-D sketch with viewer-centered representation of visible surfaces, see Jackendoff (1987, pp. 293–296).

18. To the best of my knowledge, this assertion is self-evidently true.

19. The criteria of being conscious of a bound object is left aside in what follows. While an argument could be mounted based on Young (2016) that we perceive molecular structure of odorants at the sensory level based on our ability to detect and discriminate between enantiomers with different olfactory qualities, this would oversimplify my current thinking about the object of smell and does not seem like a productive research project.

20. The section assumes that we can have conscious smell experiences as prima facie obvious, despite the lack of a dedicated corticothalamic gateway connection between olfactory sensory states and higher-level cortical olfactory processing. After all, sitting in my office at my desk sniffing a clementine with my eyes closed, not only I am both aware of the olfactory quality of a citrus fruit but this experience also has a thick phenomenology as a subjective qualitative experience for me. Moreover, I am passively aware of the smell of apple blossoms wafting in through my open window. However, it should be noted that some simply do not accept this assumption and see the opposite conclusion as obvious, such that the lack of thalamic mediation should imply that we are not conscious of smells (Shepherd, 2012).

References

Aasen, S. (2018). Spatial aspects of olfactory experience. *Canadian Journal of Philosophy, 49*(8), 1041–1061. https://doi.org/10.1080/00455091.2018.1433793

Ache, B. W., & Young, J. M. (2005). Olfaction: Diverse species, conserved principles. *Neuron, 48*, 417–430.

Aksenov, A. A., Gojova, A., Zhao, W., Morgan, J. T., Sankaran, S., Sandrock, C. E., & Davis, C. E. (2012). Characterization of volatile organic compounds in human leukocyte antigen heterologous expression systems. *Chembiochem, 13*, 1053–1059. https://doi.org/10.1002/cbic.201200011

Albano, J. E., & Wurtz, R. H. (1978). Modification of the pattern of saccadic eye movements following ablation of the monkey superior colliculus. *Neuroscience Abstracts, 4*, 161.

Algom, D., & Cain, W. S. (1991). Remembered odors and mental mixtures: Tapping reservoirs of olfactory knowledge. *Journal of Experimental Psychology: Human Perception and Performance, 17*, 1104–1119.

Allison, A. (1954). The secondary olfactory areas in the human brain. *Journal of Anatomy, 88*, 481–488.

Allport, D. A. (1987). Selection for action: Some behavioral and neurophysiological considerations of attention and action. In H. Heuer & A. F. Sanders (Eds.), *Perspectives on perception and action* (pp. 395–419). Erlbaum.

Araneda, R. C., Kini, A. D., & Firestein, S. (2000). The molecular receptive range of an odorant receptor. *Nature Neuroscience, 3*, 1248–1255.

Arctander, S. (1994). *Perfume and flavor chemicals*. Allured Publishing.

Arguedas, D., Langdon, R., & Stevenson, R. (2012a). Neuropsychological characteristics associated with olfactory hallucinations in schizophrenia. *Journal of the International Neuropsychological Society, 18*, 799–808.

Arguedas, D., Stevenson, R., & Langdon, R. (2012b). Source monitoring and olfactory hallucinations in schizophrenia. *Journal of Abnormal Psychology, 121*(4), 936–943.

Aristotle. (1984a). De Anima. In J. Barnes (Ed.), *The complete works of Aristotle* (Vol. 1, pp. 641–692). Princeton University Press.

Aristotle (1984b). De Sensu. In J. Barnes (Ed.), *The complete works of Aristotle* (Vol. 1, pp. 693–713). Princeton University Press.

Arnold, T. C., You, Y., Ding, M., Zuo, X. N., de Araujo, I., & Li, W. (2020). Functional connectome analyses reveal the human olfactory network organization. *eNeuro, 7*(4), ENEURO.0551–19.2020. https://doi.org/10.1523/ENEURO.0551-19.2020

Arshamian, A., & Larsson, M. (2014). Same same but different: The case of olfactory imagery. *Frontiers in Psychology, 5*, 34.

Arshamian, A., Olofsson, J. K., Jönsson, F. U., & Larsson, M. (2008). Sniff your way to clarity: The case of olfactory imagery. *Chemosensory Perception, 1*, 242–246. https://doi.org/10.1007/s12078-008-9035-z

Asahina, K., Louis, M., Piccinotti, S., & Vosshall, L. B. (2009). A circuit supporting concentration-invariant odor perception in *Drosophila*. *Journal of Biology, 8*, 9. https://doi.org/10.1186/jbiol108

Asai, H., Udaka, F., Hirano, M., & Ueno, S. (2008). Odor abnormalities caused by bilateralthalamic infarction. *Clinical Neurology and Neurosurgery, 110*, 500–501.

Auffarth, B. (2013). Understanding smell: The olfactory stimulus problem. *Neuroscience and Biobehavioral Reviews, 37*(8), 1667–1679.

Auvray, M., & Spence, C. (2008). The multisensory perception of flavor. *Consciousness and Cognition, 17*, 1016–1031.

Baars, B. J. (1988). *A cognitive theory of consciousness*. Cambridge University Press.

Baars, B. J. (1997). In the theatre of consciousness. *Journal of Consciousness Studies, 4*(4), 292–309.

Baars, B. J. (2002). The conscious access hypothesis: Origins and recent evidence. *Trends in Cognitive Sciences, 6*(1), 47–53.

Baars, B. J. (2013). Multiple sources of conscious odor integration and propagation in olfactory cortex. *Frontiers in Psychology, 4*, 930. https://doi.org/10.3389/fpsyg.2013.00930

Baars, B.J., Ramsøy, T.Z. & Laureys, S. (2003). Brain, conscious experience and the observing self, *TRENDS in Neurosciences*, 26, pp. 671–676.

Ballester, J., Patris, B., Symoneaux, R., & Valentin, D. (2008). Conceptual vs. perceptual wine spaces: Does expertise matter? *Food Quality and Preference, 19*, 267–276. https://doi.org/10.1016/j.foodqual.2007.08.001

Barabasz, A. F., & Gregson, R. A. (1979). Analytic wintering-over, suggestion and transient olfactory stimulation/EEG evoked potential and electrodermal responses. *Biological Psychology, 9*, 285–295.

Barnes, D. C., Hofacer, R. D., Zaman, A. R., Rennaker, R. L., & Wilson, D. A. (2008). Olfactory perceptual stability and discrimination. *Nature Neuroscience, 11*, 1378–1380.

Barwich, A. (2014). A sense so rare: Measuring olfactory experiences and making a case for a process perspective on sensory perception. *Biological Theory, 9*(3), 258–268.

Barwich, A. (2016). Making sense of smell. *The Philosophers' Magazine, 73*, 41–47.

Barwich, A. (2019). A critique of olfactory objects. *Frontiers in Psychology, 10*, 1337. https://doi.org/10.3389/fpsyg.2019.01337

Barwich, A. (2020). *Smellosophy*. Harvard University Press.

Batty, C. E. (2010a). What the nose doesn't know: Non-veridicality in olfactory experience. *Journal of Consciousness Studies*, 17, 10–17.

Batty, C. E. (2010b). Scents and sensibilia. *American Philosophical Quarterly, 47*(2), 103118.

Batty, C. E. (2010c). A representational account of olfactory experience. *Canadian Journal of Philosophy, 40*(4), 511–538.

Batty, C. E. (2014). The illusion confusion. *Frontiers in Psychology*, 5, 231.

Batty, C. E. (2015). Olfactory objects. In S. Biggs, M. Matthen, & D. Stokes (Eds.), *Perception and its modalities* (pp. 222–246). Oxford University Press.

Batty, C. E., & Young, B. D. (forthcoming). Olfactory perception. *Standford encyclopedia of philosophy*.

Beauchamp, G. K., Yamazaki, K., Wysocki, C. J., Slotnick, B. M., Thomas, L., & Boyse, E. A. (1985). Chemosensory recognition of mouse major histocompatibility types by another species. *Proceedings of the National Academy of Sciences of the United States of America, 82*, 4186–4188.

Bende, M., & Nordin, S. (1997). Perceptual learning in olfaction: Professional wine tasters versus controls. *Physiology and Behavior, 62*, 1065–1070. https://doi.org/10.1016/S0031-9384(97)00251-5

Bensafi, M., Porter, J., Pouliot, S., Mainland, J., Johnson, B., Zelano, C., Young, N., Bremner, E., Aframian, D., Khan, R., & Sobel, N. (2003). Olfactomotor activity during imagery mimics that during perception. *Nature Neuroscience, 6*, 1142–1144.

Bensafi, M., Pouliot, S., & Sobel, N. (2005). Odorant-specific patterns of sniffing during imagery distinguish "bad" and "good" olfactory imagers. *Chemical Senses, 30*, 521–552. https://doi.org/10.1093/chemse/bji045

Bensafi, M., & Rouby, C. (2007). Individual differences in odor imaging ability reflect differences in olfactory and emotional perception. *Chemical Senses, 32*, 237–244.

Bensafi, M., Rouby, C., Farget, V., Bertrand, B., Vigouroux, M., & Holley, A. (2002). Autonomic nervous system responses to odours: The role of pleasantness and arousal. *Chemical Senses, 27*, 703–709. https://doi.org/10.1093/chemse/27.8.703

Bensafi, M., Sobel, N., & Khan, R. M. (2007). Hedonic-specific activity in piriform cortex during odor imagery mimics that during odor perception. *Journal of Neurophysiology, 98*, 3254–3262.

Bensafi, M., Zelano, C., Johnson, B., Mainland, J., & Sobel, N. (2004). Olfaction: From sniff to percept. In M. Gazzaniga (Ed.), *The cognitive neurosciences III* (pp. 259–281). MIT Press.

Berger, J. (2022). Types of consciousness. In B. D. Young & C. Dicey-Jennings (Eds.), *Mind, cognition, and neuroscience*. (pp. 251–265) Routledge.

Berglund, B., Berglund, U., Lindvall, T., & Svensson, L. T. (1973). A quantitative principle of perceived intensity summation in odor mixtures. *Journal of Experimental Psychology, 100*(1), 29–38.

Bermudez, J. L. (2007). What is at stake in the debate on nonconceptual content? *Philosophical Perspectives, 21*, 55–72.

Bierling, A. L., Cory, I., Hummel, T., Cuniberti, G., & Croy, A. (2021). Olfactory perception in relation to the physicochemical odor space. *Brain Sciences, 11*(5), 563. https://doi.org/10.3390/brainsci11050563

Binder, J. R. (2015). In defense of abstract conceptual representations. *Psychonomic Bulletin and Review, 23*, 1096–1108. https://doi.org/10.3758/s13423-015-0909-1

Blankenship, M.L., Grigorova, M, & Katz, D.B. (2019. Retronasal Odor Perception Requires Taste Cortex, but Orthonasal Does Not, *Current Biology, 29*, 1–8. https://doi.org/10.1016/j.cub.2018.11.011

Blazing, R. M., & Franks, K. M. (2020). Odor coding in piriform cortex: Mechanistic insights into distributed coding. *Current Opinion in Neurobiology, 64*, 1–7. https://doi.org/10.1016/j.conb.2020.03.001

Block, N. (1993). Review of Daniel Dennett's consciousness explained. *Journal of Philosophy, 90*, 181–193.

Block, N. (1995). On a confusion about a function of consciousness. *Behavioral and Brain Sciences, 18*(2), 227–247.

Block, N. (1996). Mental paint and mental latex. In E. Villanueva (Ed.), *Perception, philosophical issues* (Vol. 7, pp. 19–49). Ridgeview.

Block, N. (2001). Paradox and cross-purposes in recent work on consciousness. In N. Block (Ed.), *Consciousness, function, and representation: Collected papers* (Vol. 1). MIT Press. Reprinted.

Block, N. (2007). Consciousness, accessibility, and the mesh between psychology and neuroscience. *Behavioral and Brain Sciences, 30*, 481–548.

Block, N. (2008). Consciousness and cognitive access. *Proceedings of the Aristotelian Society, 108*, 289–317. https://doi.org/10.1111/j.1467-9264.2008.00247.x

Block, N (2009). Comparing the major theories of consciousn ess. In M. Gazzaniga (Ed.), *The cognitive neurosciences IV*. MIT Press.pp. 1111–1123.

Block, N. (2011). Perceptual consciousness overflows cognitive access. *Trends in Cognitive Sciences, 15*, 567–575. https://doi.org/10.1016/j.tics.2011.11.001

Boelens, M. H., & van Gemert, L. J. (1993). Sensory properties of optical isomers. *Perfumer and Flavorist, 18*, 1–15.

Bojsen-Moller, F., & Fahrenkrug, J. (1971). Nasal swell-bodies and cyclic changes in the air passage of the rat and rabbit nose. *Journal of Anatomy, 110*, 25–37.

Bolding, K. A., & Franks, K. M. (2017). Complementary codes for odor identity and intensity in olfactory cortex. *Elife, 6*, e22630. https://doi.org/10.7554/eLife.22630

Boly, M., Massimini, M., Tsuchiya, N., Postle, B. R., Koch, C., & Tononi, G. (2017). Are the neural correlates of consciousness in the front or in the back of the cerebral cortex? Clinical and Neuroimaging Evidence. *Journal of Neuroscience, 37*, 9603–9613. https://doi.org/10.1523/JNEUROSCI.3218-16.2017

Boyle, J. A., Djordjevic, J., Olsson, M. J., Lundström, J. N., & Jones-Gotman, M. (2009). The human brain distinguishes between single odorants and binary mixtures. *Cerebral Cortex, 19*, 66–71. https://doi.org/10.1093/cercor/bhn058

Bozza, T. C., & Mombaerts, P. (2001). Olfactory coding: Revealing intrinsic representations of odors. *Current Biology, 11*(17), R687–R690.

Brooks, R. A. (1994). Coherent behaviour from many adaptive processes. In D. Cliff, P. Husbands, J.-A. Meyer, & S. W. Wilson (Eds.), *From animals to animats 3: Proceedings of the Third International Conference on Simulation of Adaptive Behavior* (pp. 22–29). MIT Press.

Buck, L. B., & Axel, R. (1991). A novel multigene family may encode odorant receptors: A molecular basis for odor recognition. *Cell, 65*, 175–189.

Budek, T., & Farkas, K. (2014). Which causes of an experience are also objects of the experience? In B. Brogaard (Ed.), *Does perception have content?* (pp. 351–370). Oxford University Press.

Burge, T. (2010). *Origins of objectivity*. Oxford University Press.

Bushdid, C., Magnasco, M. O., Vosshall, L. B., & Keller, A. (2014). Humans can discriminate more than 1 trillion olfactory stimuli. *Science, 343*, 1370–1372.

Cain, W. S. (1979). Know with the nose: Keys to odor identification. *Science, 203*, 467–470.

Cain, W. S., Stevens, J. C., Nickou, C. M., Giles, A., Johnston, I., & Garcia-Medina, M. R. (1995). Life-span development of odor identification, learning, and olfactory sensitivity. *Perception, 24*, 1457–1472.

Carello, C. D., & Krauzlis, R. J. (2004). Manipulating intent: Evidence for a causal role of the superior colliculus in target selection. *Neuron, 43*(4), 575–583.

Carmichael, S. T., Clugnet, M. C., & Price, J. L. (1994). Central olfactory connections in the macaque monkey. *Journal of Comparative Neurology, 346*, 403–434.

Carvalho, F. (2014). Olfactory objects. *Disputatio, 6*(38), 45–66.

Casagrande, V. A., & Diamond, I. T. (1974). Ablation study of the superior colliculus in the tree shrew (*Tupaia glis*). *Journal of Comparative Neurology, 156*, 207–238.

Castro, J. B., & Seeley, W. P. (2014). Olfaction, valuation, and action: Reorienting perception. *Frontiers in Psychology, 5*, e73289.

Cavedon-Taylor, D. (2018). Odors, objects and olfaction. *American Philosophical Quarterly, 55*(1), 81–94.

Cavelius, M., Brunel, T., & Didier, A. (2022). Lessons from behavioral lateralization in olfaction. *Brain Structure and Function, 227*(2), 685–696. https://doi.org/10.1007/s00429-021-02390-w

Chaix, R., Cao, C., & Donnelly, P. (2008). Is mate choice in humans MHC-dependent? *PLoS Genetics, 4*, e1000184. https://doi.org/10.1371/journal.pgen.1000184

Chan, K. Q., van Dooren, R., Holland, R. W., & van Knippenberg, A. (2019). Disgust lowers olfactory threshold: A test of the underlying mechanism. *Cognition and Emotion, 34*(3), 621–627. https://doi.org/10.1080/02699931.2019.1660145

Chen, C., Zou, D., Altomare, C. G., Xu, L., Greer, C. A., & Firestein, S. J. (2014). Nonsensory target-dependent organization of piriform cortex. *Proceedings of the National Academy of Sciences of the United States of America, 111*, 16931–16936. https://doi.org/10.1073/pnas.1411266111

Chrea, C., Valentin, D., Sulmont-Rossé, C., Ly Mai, H., Nguyen, D. H., & Abdi, H. (2005a). Culture and odor categorization: Agreement between cultures depends upon the odors. *Food Quality and Preference, 15*, 669–679. https://doi.org/10.1016/j.foodqual.2003.10.005

Chrea, C., Valentin, D., Sulmont-Rossé, C., Ly Mai, H., Nguyen, D. H., & Abdi, H. (2005b). Semantic, typicality and odor representation. *Chemical Senses, 30*, 37–49. https://doi.org/10.1093/chemse/bjh255

Clark, A. (1993). *Sensory qualities*. Clarendon Press.

Clarke, A., Devereux, B. J., Randall, B., & Tyler, L. K. (2015). Predicting the time course of individual objects with MEG. *Cerebral Cortex, 25*, 3602–3612. https://doi.org/10.1093/cercor/bhu203

Clarke, A., Taylor, K. I., Devereux, B., Randall, B. & Tyler, L. K. (2013). From perception to conception: How meaningful objects are processed over time. Cerebral Cortex, 23, 187–197.

Clarke, A., & Tyler, L. K. (2014). Object-specific semantic coding in human perirhinal cortex. *Journal of Neuroscience, 34*, 4766–4775.

Cleland, T. A., Chen, S. T., Hozer, K. W., Ukatu, H. N., Wong, K. J., & Zheng, F. (2012). Sequential mechanisms underlying concentration invariance in biological olfaction. *Frontiers in Neuroengineering, 4*, 21.

Cleland, T. A., & Linster, C. (2019). Central olfactory structures. In R. L. Doty (Ed.), *Handbook of clinical neurology* (Vol. 164, 3rd series, Smell and taste, pp. 79–96). Elsevier.

Comanski, B. (2020). Spatial experience and olfaction: A role for naïve topology. *Mind and Language, 37*(4), 715–733. https://doi.org/10.1111/mila.12354

Connell, L., & Lynott, D. (2014). Principles of representation: Why you can't represent the same concept twice. *Topics in Cognitive Science, 6*, 390–406.

Courtiol, E., & Wilson, D. A. (2014). Thalamic olfaction: Characterizing odor processing in the mediodorsal thalamus of the rat. *Journal of Neurophysiology, 111*, 1274–1285. https://doi.org/10.1152/jn.00741.2013

Courtiol, E., & Wilson, D. A. (2016a). Neural representation of odor-guided behavior in the rat olfactory thalamus. *Journal of Neuroscience, 36*, 5946–5960. https://doi.org/10.1523/JNEUROSCI.0533-16.2016

Couritol, E., & Wilson, D. A. (2016b). The olfactory mosaic—Bringing an olfactory network together for odor perception. *Perception, 46*(3–4), 320–332. https://doi.org/10.1177/0301006616663216

Cox, R. E., & Langdon, R. A. (2016). Hypnotic olfactory hallucinations. *International Journal of Clinical and Experimental Hypnosis, 64*(1), 24–44. https://doi.org/10.1080/00207144.2015.1099401

Crane, T. (1990). The language of thought: No syntax without semantics. *Mind and Language, 5*, 187–212.

Crick, F. (1984). Function of the thalamic reticular complex: The searchlight hypothesis. *Proceedings of the National Academy of Sciences of the United States of America, 81*, 4586–4590.

Crick, F. (1994). *The astonishing hypothesis*. Scribners.

Crick, F., & Koch, C. (1990). Towards a neurobiological theory of consciousness. *The Neurosciences, 2*, 263–275.

Crick, F., & Koch, C. (1998). Consciousness and neuroscience. *Cerebral Cortex, 8*(2), 97–107.

Crick, F., & Koch, C. (2003). A framework for consciousness. *Nature Neuroscience, 6*(2), 119–127.

Crick, F., & Koch, C. (2005). What is the function of the claustrum? *Philosophical Transactions Biological Sciences, 360*, 1271–1279.

Croijmans, I., Arshamian, A., Speed, L. J., & Majid, A. (2021). Wine experts' recognition of wine odors is not verbally mediated. *Journal of Experimental Psychology, 150*(3), 545–559. https://doi.org/10.1037/xge0000949

Croijmans, I., & Majid, A. (2016). Not all flavor expertise is equal: The language of wine and coffee experts. *PLoS One, 11*(6), e0155845. https://doi.org/10.1371/journal.pone.0155845

Croijmans, I., Speed, L. J., Arshamian, A., & Majid, A. (2020). Expertise shapes multimodal imagery for wine. *Cognitive Science, 44*, e12842. https://doi.org/10.1111/cogs.12842

Cross, D. J., Flexman, J. A., Anzai, Y., Morrow, T. J., Maravilla, K. R., & Minoshima, S. (2006). In vivo imaging of functional disruption, recovery and alteration in rat olfactory circuitry after lesion. *Neuroimage, 32*, 1265–1272.

Cummings, D. M., & Belluscio, L. (2008). Charting plasticity in the regenerating maps of mammalian olfactory bulb. *The Neuroscientist, 14*(3), 251–263.

Cummins, R. (1996). Systematicity. *Journal of Philosophy, 93*, 591–614.

Cummins, R., Blackmon, J. Byrd, D., Poirier. P. Roth, M., & Schwarz, G. (2001). Systematicity and the cognition of structured domains. *Journal of Philosophy, 98*, 167–185.

Dalal, T., Gupta, N., & Haddad, R. (2020). Bilateral and unilateral odor processing and odor perception. *Communications Biology, 3*, 150. https://doi.org/10.1038/s42003-020-0876-6

Dean, P., Redgrave, P., & Westby, G. M. W. (1989). Event or emergency—2 response systems in the mammalian superior colliculus. *Trends in Neurosciences, 12*(1), 37–47.

Deems, D. A., Doty, R. L., Settle, R. G., Shaman, P., Mester, A. F., Kimmelman, C. P., Brightman, V. J., & Snow, J. B. (1991). Smell and taste disorders: A study of 750 patients from the University of Pennsylvania Smell and Taste Center. *Archives of Otolaryngology—Head and Neck Surgery, 117*(5), 519–528.

Dehaene, S., & Changeux, J. (2003). Neural mechanisms for access to consciousness. In M. Gazzainiga (Ed.), *The cognitive neurosciences*. MIT Press. pp. 1145–1157

Dehaene, S., Changeux, J., Naccache, L., Sackur, J., & Sergent, C. (2006). Consciousness, preconscious, and subliminal processing: A testable taxonomy. *Trends in Cognitive Sciences, 10*(5), 204–211.

Dehaene, S., & Naccache, L. (2001). Towards a cognitive neuroscience of consciousness. *Cognition, 79*, 1–37.

Del Pinal, G. (2015). Prototypes as compositional components of concepts. *Synthese, 193*, 2899–2927. https://doi.org/10.1007/s11229-015-0892-0

Denny-Brown, D. (1962). The midbrain and motor integration. *Proceedings of the Royal Society of Medicine, 55*, 527–538.

de Olmos, J., Hardy, H., & Heimer, L. (1978). The afferent connections of the main and the accessory olfactory bulb formations in the rat: An experimental HRP study. *Journal of Comparative Neurology, 15*, 213–244.

DePay, K. S., & Hornung, D. E. (2002). The contribution each nostril makes to olfactory perception. Presented at the 24th Annual Conference of the Association for Chemoreception Sciences.

Deroy, O. (2009). The power of tastes reconciling science and subjectivity. In B. Smith (Ed.), *Questions of taste*. (pp. 99–126) Oxford University Press.

Deroy, O. (2019). Categorising without concepts. *Review of Philosophy and Psychology, 10*, 465–478. https://doi.org/10.1007/s13164-019-00431-2

Derti, A., Cenik, C., Kraft, P., & Roth, F. P. (2010). Absence of evidence for MHC-dependent mate selection within HapMap populations. *PLoS Genetics, 6*, e1000925. https://doi.org/10.1371/journal.pgen.1000925

Derti, A., & Roth, F. P. (2013). Response to "MHC-dependent mate choice in humans: Why genomic patterns from the HapMap European American data set support the hypothesis." *Bioessays, 34*, 576–577. https://doi.org/10.1002/bies.201200023

de Sousa, H. (2011). Changes in the language of perception in Cantonese. *The Senses and Society, 6*(1), 38–47. https://doi.org/ 10.2752/174589311X12893982233678

Dever, J. (2006). Compositionality. In E. Lepore & B. Smith (Eds.), *The Oxford handbook of philosophy of language* (pp. 633–666). Oxford University Press.

Dever, J. (2012). Compositionality. In *The Routledge handbook to the philosophy of language* (pp. 91–102). Taylor & Francis.

Djordjevic, J., Zatorre, R. J., Petrides, M., Boyle, J. A., & Jones-Gotman, M. (2005). Functional neuroimaging of odor imagery. *Neuroimage, 24*, 791–801. https://doi.org/10.1016/j.neuroimage.2004.09.035

Djordjevic, J., Zatorre, R., Petrides, M., & Jones-Gotman, M. (2004). The mind's nose: Effects of odor and visual imagery on odor detection. *Psychological Science, 15*, 143–148. https://doi.org/10.1111/j.0956-7976.2004.01503001.x

Doerig, A., Schurger, A., & Herzog, M. H. (2021). Hard criteria for empirical theories of consciousness. *Cognitive Neuroscience, 12*(2), 41–62. https://doi.org/10.1080/17588928.2020.1772214

Dorrego-Rivas, A., & Grubb, M. S. (2022). Developing and maintaining a nose-to-brain map of odorant identity. *Open Biology, 12*, 220053. https://doi.org/10.1098/rsob.220053

Doty, R. L. (1995). Intranasal trigeminal chemoreception: Anatomy, physiology, and psychophysics. In R. L. Doty (Ed.), *Handbook of olfaction and gustation* (pp. 821–883). Marcel Dekker.

Doty, R. L. (2010). *The great pheromone myth*. Johns Hopkins University Press.

Dou, W., Li, Y., Geisler, M. W., & Morsella, E. (2018). Involuntary polymodal imagery involving olfaction, audition, touch, taste, and vision. *Consciousness and Cognition, 62*, 9–20. https///doi.org/10.1016/j.concog.2018.04.007

Dretske, F. (1981). *Knowledge and the flow of information*. MIT Press.

Dulac, C. (2000). Sensory coding of pheromone signals in mammals. *Current Opinions in Neurobiology, 10*, 511–518.

Dulay, M. F., Gesteland, R. C., Shear, P. K., Ritchey, P. N., & Frank, R. A. (2008). Assessment of the influence of cognition and cognitive processing speed on three tests of olfaction. *Journal of Clinical and Experimental Neuropsychology, 30*(3), 327–337.

Earley, J. E. (2005). Why there is no salt in the sea. *Foundations of Chemistry, 7*, 85–102.

Echevarria-Cooper, S. L., Zhou, G., Zelano, C., Pestilli, F., Parrish, T. B., & Kahnt, T. (2022). Mapping the microstructure and striae of the human olfactory tract with diffusion MRI. *Journal of Neuroscience, 42*(1), 58–68. https://doi.org/10.1523/JNEUROSCI.1552-21.2021

Ehman, K. D., & Scott, E. (2001). Urinary odour preferences of MHC congenic female mice, *Mus domesticus*: Implications for kin recognition and detection of parasitized males. *Animal Behaviour, 62*, 781–789. https://doi.org/10.1006/anbe.2001.1805

Eichenbaum, H., Shedlack, K. J., & Eckmann, K. W. (1980). Thalamocortical mechanisms in odor-guided behavior. I. Effects of lesions of the mediodorsal thalamic nucleus and frontal cortex on olfactory discrimination in the rat. *Brain, Behavior and Evolution, 17*, 255–275.

Engen, T., & Ross, B. M. (1973). Long-term memory of odors with and without verbal descriptors. *Journal of Experimental Psychology, 100*, 221–227.

Fallon, A. E., & Rozin, P. (1983). The psychological bases of food rejections by humans. *Ecology of Food and Nutrition, 13*, 15–26.

Filiou, R., Lepore, F., Bryant, B., Lundström, J. N., & Frasnelli, J. (2015). Perception of trigeminal mixtures. *Chemical Senses, 40*, 61–69.

Firestein, S. (2001). How the olfactory system makes sense of scents. *Nature, 413*, 211–218.

Firestein, S., & Werblin, F. (1989). Odor-induced membrane currents in vertebrate-olfactory receptor neurons. *Science, 244*, 79–82.

Fodor, J. A. (1976). *The language of thought*. The Harvester Press.

Fodor, J. A. (1981). Methodological solipsism considered as a research strategy in cognitive psychology. In J. Fodor (Ed.), *Representations* (pp. 225–253). The Harvester Press.

Fodor, J. A. (1985). Fodor's Guide to Mental Representation: The Intelligent Auntie's Vade-Mecum. Mind 94 (373):76–100.

Fodor, J. A. (1987). *Psychosemantics*. MIT Press.

Fodor, J. A. (1990). Psychosemantics or: Where do truth conditions come from? In W. G. Lycan (Ed.), *Mind and cognition* (1st ed.). Blackwell Publishers.

Fodor, J. A. (2000). *The mind doesn't work that way*. MIT Press.

Fodor, J. A. (2008). *LOT2*. Oxford University Press.

Fodor, J. A., & Lepore, E. (1992). *Holism: A shopper's guide*. Blackwell Publishers.

Fodor, J., & McLaughlin, B. P. (1990). Connectionism and the problem of systematicity. *Cognition, 35*, 83–204.

Fodor, J. A., & Pylyshyn, A. (1988). Connectionism and cognitive architecture: A critical analysis. In S. Pinker & J. Mehler (Eds.), *Connections and symbols* (Vol. 28, pp. 3–71). MIT Press.

Frank, R. A., Dulay, M. F., & Gesteland, R. C. (2003). Assessment of the sniff magnitude test as a clinical test of olfactory function. *Physiology and Behavior, 78*, 195–204. https://doi.org/10.1016/S0031-9384(02)00965-4

Fredericksen, K. E., McQueen, K. A., & Samuelsen, C. L. (2019). Experience-dependent c-Fos expression in the mediodorsal thalamus varies with chemosensory modality. *Chemical Senses, 44*, 41–49. https://doi.org/10.1093/chemse/bjy070

Friedrich, R. W., & Laurent, G. (2001). Dynamic optimization of odor representation by slow temporal patterning of mitral cell activity. *Science, 291*, 889–894.

Frixione, M., & Lieto, A. (2012). Representing concepts in formal ontologies. *Logic and Logical Philosophy, 21*, 391–414. https://doi.org/10.12775/LLP.2012.018

Fulkerson, M. (2014). Rethinking the senses and their interactions. *Frontiers in Psychology, 5*, 1426. https://doi.org/10.3389/fpsyg.2014.01426

Garver-Apgar, C. E., Gangestad, S. W., Thornhill, R., Miller, R. D., & Olp, J. J. (2006). Major histocompatibility complex alleles, sexual responsivity, and unfaithfulness in romantic couples. *Psychological Science, 17*, 830–835. https://doi.org/10.1111/j.1467-9280.2006.01789.x

Gerkin, R. C., & Castro, J. B. (2015). The number of olfactory stimuli that humans can discriminate is still unknown. *eLife, 4*, 1–15. https://doi.org/10.7554/eLife.08127

Giaffar, H., Rinberg, D., & Koulakov, A. A. (2018). *Primacy model and the evolution of the olfactory receptor repertoire*. bioRxiv. https://doi.org/10.1101/255661

Gilad, Y., Wiebe, V., Przeworski, M., Lancet, D., & Pääbo, S. (2004). Loss of olfactory receptor genes coincides with the acquisition of full trichromatic vision in primates. *PLoS Biology, 2*(1), e5.

Gilbert, A. N. (2008). *What the nose knows: The science of scent in everyday life*. Crown Publishers.

Gilbert, A. N., Crouch, M., & Kemp, S. E. (1998). Olfactory and visual mental imagery. *Journal of Mental Imagery, 22*(3 and 4), 137–146.

Gilbert, A. N., & Rosenwasser, A. M. (1987). Biological rhythmicity of nasal airway patency: A re-examination of the "nasal cycle." *Acta Otolaryngologica, 104*, 180–186.

Giphart, M. J., & D'Amaro, J. (1983). HLA and reproduction? *International Journal of Immunogenetics,* 10, 25–29.

Goldberg, E. M., Wang, K., Goldberg, J., & Aliani, M. (2018). Factors affecting the ortho- and retronasal perception of flavors: A review. *Critical Reviews in Food Science and Nutrition, 58*(6), 913–923. https://doi.org/10.1080/10408398.2016.1231167

Goldstein, A., & Young, B. D. (2022). The unconscious mind. In B. D. Young & C. Dicey-Jennings (Eds.), *Mind, cognition, and neuroscience*. (pp. 344–364) Routledge.

Gonzalez, J., Barros-Loscertales, A., Pulvermüller, F., Meseguer, V., Sanjuán, A., Belloch, V., & Ávila, C. (2006). Reading cinnamon activates olfactory brain regions. *NeuroImage, 32*, 906–912. https://doi.org/10.1016/j.neuroimage.2006.03.037

Goodglass, H., Barton, M. I., & Kaplan, E. F. (1968). Sensory modality and object-naming in aphasia. *Journal of Speech and Hearing Research, 11*, 488–496.

Gottfried, J. A. (2010). Central mechanisms of odour object perception. *Nature Reviews Neuroscience, 11*, 628–641.

Gray, R., & Tanesini, A. (2010). Perception and action: The taste test. *Philosophical Quarterly, 60*, 718–734.

Greer, C. A., Stewart, W. B., Kauer, J. S., & Shepherd, G. M. (1981). Topographical and laminar localization of 2-deoxyglucose uptake in rat olfactory bulb induced by electrical stimulation of olfactory nerves. *Brain Research, 217*, 279–293.

Grobman, M., Dalal, T., Lavian, H., Xu, F., Korngreen, A., & Haddad, R. (2018). A mirror-symmetric excitatory link coordinates odor maps across olfactory bulbs and enables odor perceptual unity. *Neuron, 99*, 800–813. https://doi.org/10.1016/j.neuron.2018.07.012

Gross, M. (2019). Odour space—the final frontier. *Current Biology, 29*, R663–R682.

Gross-Isseroff, R., & Lancet, D. (1988). Concentration-dependent changes of perceived odor quality. *Chemical Senses, 13*, 191–204.

Haberly, L. B. (2001). Parallel-distributed processing in olfactory cortex: New insights from morphological and physiological analysis of neuronal circuitry. *Chemical Senses, 26*, 551–576.

Haddad, R., Lanjuin, A., Madisen, L., Zeng, H., Murthy, V. N., & Uchida, N. (2013). Olfactory cortical neurons read out a relative time code in the olfactory bulb. *Nature Neuroscience, 16*, 949–957.

Haddad, R., Lapid, H., Harel, D., & Sobel, N. (2008). Measuring smells. *Current Opinion in Neurobiology, 18*(4), 438–444.

Haddad, R., Weiss, T., Khan, R., Nadler, B., Mandairon, N., Bensafi, M., Schneidman, E., & Sobel, N. (2010). Global features of neural activity in the olfactory system form a parallel code that predicts olfactory behavior and perception. *Journal of Neuroscience, 30*(27), 9017–9026.

Hallem, E. A., & Carlson, J. R. (2006). Coding of odors by a receptor repertoire. *Cell, 125*, 143–160.

Hallem, E. A., Dahanukar, A., & Carlson, J. R. (2006). Insect odor and taste receptors. *Annual Review of Entomology, 51*, 113–135.

Halpern, M. (1987). The organization and function of the vomeronasal system. *Annual Review of Neuroscience, 10*, 325–362.

Harland, R. E., & Frank, R. A. (1997). Exploratory sniffing as an index of olfactory functioning. Poster Presented at the International Symposium on Olfaction and Taste. https://achems.org/web/downloads/programs/Program97.pdf

Hasegawa, M., & Kern, E. B. (1977). The human nasal cycle. *Mayo Clinical Proceedings, 52*, 28–34.

Havlicek, J., & Roberts, C. S. (2009). MHC-correlated mate choice in humans: A review. *Psychoneuroendocrinology, 34*, 497–512.

Heilmann, S., & Hummel, T. (2004). A new method for comparing orthonasal and retronasal olfaction. *Behavioral Neuroscience, 118*, 412–419.

Henkin, R. I., Levy, L. M., & Fordyce, A. (2013a). Taste and smell function in chronic disease. *American Journal of Otolaryngology, 34*, 477–489.

Henkin, R. I., Potolicchio, S. J., & Levy, L. M. (2013b). Olfactory hallucinations without clinical motor activity/A comparison of unirhinal with birhinal phantosmia. *Brain Science, 3*, 1483–1553. https://doi.org/10.3390/brainsci3041483

Herz, R. S. (2000). Verbal coding in olfactory versus nonolfactory cognition. *Memory and Cognition, 28*(5), 957–964.

Herz, R. S., & Cahill, E. D. (1997). Differential use of sensory information in sexual behavior as a function of gender. *Human Nature, 8*(3), 275–286. https://doi.org/10.1007/BF02912495

Herz, R. S., & Inzlicht, M. (2002). Sex differences in response to physical and social factors involved in human mate selection. The importance of smell for women. *Evolution and Human Behavior, 23*, 359–364. https://doi.org/10.1016/S1090-5138(02)00095-8

Herz, R. S., & Schooler, J. W. (2002). A naturalistic study of autobiographical memories evoked by olfactory and visual cues: Testing the Proustian hypothesis. *American Journal of Psychology, 115*, 21–32.

Hinds, J. W., Hinds, P. L., & McNelly, N. A. (1984). An autoradiographic study of the mouse olfactory epithelium: Evidence for long-lived receptors. *The Anatomical Record, 210*(2), 375–383.

Homewood, J., & Stevenson, R. J. (2001). Differences in naming accuracy of odors presented to the left and right nostrils. *Biological Psychology, 58*, 65–73.

Horgan, T., & Tienson, J. (1994). A nonclassical framework for cognitive science. *Synthese, 101*, 305–345.

Horgan, T., & Tienson, J. (1996). *Connectionism and the philosophy of psychology*. MIT Press.

Hornung, D. E., & Enns, M. P. (1986). Possible mechanisms for the processes of referred taste and retronasal olfaction. *Chemical Senses, 11*, 616.

Huisman, J. L. A., & Majid, A. (2018). Psycholinguistic variables matter in odor naming. *Memory and Cognition, 46*, 577–588. https://doi.org/10.3758/s13421-017-0785-1

Hummel, T. (2000). Assessment of intranasal trigeminal function. *International Journal of Psychophysiology, 36*, 147–155.

Hummel, T. (2008). Retronasal perception of odors. *Chemistry and Biodiversity, 5*(6), 853–861.

Hummel, T., Guel, H., & Delank, W. (2004). Olfactory sensitivity of subjects working in odorous environments. *Chemical Senses*, 29, 533–536. https://doi.org/10.1093/chemse/bjh058

Hummel, T., Heilmann, S., Landis, B. N., Reden, B. N., Frasnelli, J., Small, D. M., & Gerber, J. (2006). Perceptual differences between chemical stimuli presented through the ortho- or retronasal route. *Flavour and Fragrance Journal, 21*, 42–47.

Hummel, T., & Livermore, A. (2002). Intranasal chemosensory function of the trigeminal nerve and aspects of its relation to olfaction. *International Archives of Occupational and Environmental Health, 75*(5), 305–313.

Humphries, C., Binder, J. R., Medler, D. A., & Liebenthal, E. (2006). Syntactic and semantic modulation of neural activity during auditory sentence comprehension. *Journal of Cognitive Neuroscience, 18*, 665–679.

Hym, C., Forma, V., Anderson, D. I., Provasi, J., Granjon, L., Huet, V., Carpe, E., Teulier, C., Durand, K., Schaal, B., & Barbu-Roth, M. (2020). Newborn crawling and rooting in response to maternal breast odor. *Developmental Science, 24*, e13061. https://doi.org/10.1111/desc.13061

Ihara, Y., Aoki, K., Tokunaga, K., Takahashi, K., & Juji, T. (2000). HLA and human mate choice: Tests on Japanese couples. *Anthropological Science, 108*, 199–214. https://doi.org/10.1537/ase.108.199

Ijichi, C., Wakabayashi, H., Sugiyama, S., Ihara, Y., Nogi, Y., Nagashima, A., Ihara, S., Niimura, Y., Shimizu, Y., Kondo, K., & Touhara, K. (2019). Metabolism of odorant molecules in human nasal/oral cavity affects the odorant perception. *Chem Senses, 44*(7), 465–481. https://doi.org/10.1093/chemse/bjz041.

Illig, K. R., & Haberly, L. B. (2003). Odor-evoked activity is spatially distributed in piriform cortex. *Journal of Comparative Neurology, 457*, 361–373.

Imamura, K., Mataga, N., & Mori, K. (1992). Coding of odor molecules by mitral/tufted cells in rabbit olfactory bulb. I. Aliphatic compounds. *Journal of Neurophysiology, 68*(6), 1986–2002.

Isa, T., & Kobayashi, Y. (2004). Switching between cortical and subcortical sensorimotor pathways. *Progress in Brain Research, 143*, 299–305.

Jackendoff, R. (1987). *Consciousness and the computational mind.* MIT Press.

Jackson, F. (1977). *Perception: A representative theory*. Cambridge University Press.

Jacob, S., McClintock, M. K., Zelano, B., & Ober, C. (2002). Paternally inherited HLA alleles are associates with women's choice of male odor. *Nature Genetics, 30*, 175–179.

James, A. (2018). How Robert Parker's 90 and Ann Noble's aroma wheel changed the discourse of wine tasting notes. *ILCEA*, 31.

Jehl, C., Royet, J. P., & Holley, A. (1997). Role of verbal encoding in short- and long-term odor recognition. *Perception and Psychophysics, 59*, 100–110.

Jellinek, P. (1951). *Die Psychologischen Grundlagen der Parfumerie Untersuchungen uber die Wirkungen von Geruchen auf das Gefuhlsleben.* Dr. Alfred Hüthig.

Jin, K., Speed, T. P., & Thomson, G. (1995). Tests of random mating for a highly polymorphic locus—Application to HLA data. *Biometrics, 51*, 1064–1076. https://doi.org/10.2307/2533005

Johnson, B. A., & Leon, M. (2007). Chemotopic odorant coding in a mammalian olfactory system. *Journal of Computational Neurology, 503*, 1–34.

Johnson, B. A., Ong, J., & Leon, M. (2010). Glomerular activity patterns evoked by natural odor objects in the rat olfactory bulb are related to patterns evoked by major odorant components. *Journal of Comparative Neurology, 518*(9), 1542–1555. https://doi.org/10.1002/cne.22289

Jönsson, F. U., Olosson, H., & Olsson, M. J. (2005). Odor emotionally affects the confidence in odor naming. *Chemical Senses, 30*, 29–35.

Jraissati, Y., & Deroy, O. (2021). Categorizing smells: A localist approach. *Cognitive Science, 45*, e12930. https://doi.org/10.1111/cogs.12930

Jylkka, J. (2011). Hybrid extensional prototype compositionality. *Minds and Machines, 21*, 41–56. https://doi.org/10.1007/s11023-010-9217-8

Kamimura, S., Masaoka, Y., Yoshikawa, A., Kamijo, S., Ohtaki, H., Koiwa, N., Honma, M., Sakikawa, K., Kobayashi, H., & Izumizaki, M. (2022). New granule cells in the olfactory bulb are associated with high respiratory input in an enriched odor environment. *Neuroscience Research, 182*, 52–59. https://doi.org/10.1016/j.neures.2022.05.007

Kareken, D. A., Sabri, M., Radnovich, A., Claus, E., Foresman, B., Hector, D., & Hutchins, G. D. (2004). Olfactory system activation from sniffing: Effects in piriform and orbitofrontal cortex. *NeuroImage, 22*(1), 456–465.

Kärnekull, S. C., Jönsson, F. U., Willander, J., Sikström, S., & Larsson, M. (2015). Long-term memory for odors—Influences of familiarity and identification across 64 days. *Chemical Senses, 40*(4), 259–267. https://doi.org/10.1093/chemse/bjv003

Kay, L. M., Crk, T., & Thorngate, J. (2005). A redefinition of odor mixture quality. *Behavioral Neuroscience, 119*(3), 726–733.

Kay, L. M., & Sherman, S. M. (2006). An argument for an olfactory thalamus. *Trends in Neuroscience, 30*(2), 47–53.

Keller, A. (2017). *Philosophy of olfactory perception.* Palgrave Macmillan.

Keller, A., & Malaspina, D. (2013). Hidden consequences of olfactory dysfunction: A patient report series. *BMC Ear, Nose and Throat Disorders, 13*, 8. https://doi.org/10.1186/1472-6815-13-8

Keller, A., & Vosshall, L. B. (2016). Olfactory perception of chemically diverse molecules. *BMC Neuroscience, 17*, 55.

Keller, A., & Young, B. D. (2023). *Theoretical perspectives on smell*. Routledge.

Kepecs, A., Uchida, N., & Mainen, Z. F. (2006). The sniff as a unit of olfactory processing. *Chemical Senses, 31*, 167–179.

Kermen, F., Chakirian, A., Sezille, C., Joussain, P., Le Goff, G., Ziessel, A., Chastrette, M., Mandairon, N., Didier, A., Rouby, C., & Bensafi, M. (2011). Molecular complexity determines the number of olfactory notes and the pleasantness of smells. *Scientific Reports, 1*, 206.

Keverne, E. B. (1999). The vomeronasal organ. *Science, 286*, 716–720.

Keyhani, K., Scherer, P. W., & Mozell, M. M. (1997). A numerical model of nasal odorant transport for the analysis of human olfaction. *Journal of Theoretical Biology, 186*, 279–301.

Kleemann, A. M., Albrecht, J., Schöpf, V., Haegler, K., Kopietz, R., Hempel, J. M., Linn, J., Flanagin, V. L., Fesl, G., & Wiesmann, M. (2009). Trigeminal perception is necessary to localize odors. *Physiology and Behavior, 97*, 401–405. https://doi.org/10.1016/j.physbeh.2009.03.013

Klopping, H. L. (1971). Olfactory theories and the odors of small molecules. *Journal of Agricultural and Food Chemistry, 19*(5), 999–1004.

Koch, C. (2004). *The quest for consciousness: A neurobiological approach*. Roberts and Co.

Koch, C., Massimini, M., Boly, M., & Tononi, G. (2016). Neural correlates of consciousness: Progress and problems. *Nature Reviews Neuroscience, 17*(5), 307–321. https://doi.org/10.1038/nrn.2016.22

Koritnik, B., Azam, S., Andrew, C. M., Leigh, O. N., & Williams, S. C. (2008). Imaging the brain during sniffing: A pilot fMRI study. *Pulmonary Pharmacology and Therapeutics, 22*(2), 97–101.

Kosslyn, S. M. (2003). Understanding the mind's eye . . . and nose. *Nature Neuroscience, 6*(11), 1124–1125.

Kosslyn, S. M., Ganis, G., & Thompson, W. L. (2003). Mental imagery: Against the nihilistic hypothesis. *Trends in Cognitive Sciences, 7*(3), 109–111.

Kouider, S., de Gardelle, V., & Sackur, J. (2012). Do we still need phenomenal consciousness? Comment on block. *Trends in Cognitive Sciences, 16*, 140–141. https://doi.org/10.1016/j.tics.2012.01.003

Kratskin, I. L., & Belluzzi, O. (2003). Anatomy and neurochemistry of the olfactory bulb. In R. L. Doty (Ed.), *Handbook of olfaction and gustation* (2nd ed., pp. 139–164). Marcel Dekker.

Kronenbuerger, M., Zobel, S., Igner, J., Finkelmeyer, A., Reinacher, P., Coenen, V. A., Wilms, H., Kloss, M., Kienig, K., Daniels, C., Falk, D., Schulz, J. B., Deushci, G., & Hummel, T. (2010). Effects of deep brain stimulation of the cerebellothalamic pathways on the sense of smell. *Experimental Neurology, 222*, 142–152.

Kumar, R., Kaur, R., Auffarth, B., & Bhondekar, A. P. (2015). Understanding the odour spaces: A step towards solving olfactory stimulus-percept problem. *PLoS One, 10*(10), e0141263.

Laing, D. G., & Francis, G. W. (1989). The capacity of humans to identify odors in mixtures. *Physiology and Behavior, 46*(5), 809–814.

Laing, D. G., & Glemarec, A. (1992). Selective attention and the perceptual analysis of odor mixtures. *Physiology and Behavior, 52*, 1047–1053.

Laing, D. G., & Jinks, A. (1999). Odor identification in mixtures: Is olfactory working memory the ultimate limitation? *Chemical Senses, 24*(5), 583.

Laing, D. G., Jinks, A., Link, C., & Hutchinson, I. (2001). The capacity of humans to analyse odor mixtures and taste mixtures is limited by working memory. *Chemical Senses, 26*(6), 702.

Laing, D. G., Link, C., Jinks, A. L., & Hutchinson, I. (2002). The limited capacity of humans to identify the components of taste mixtures and taste—odor mixtures. *Perception, 31*(5), 617–635.

Lamme, V. A. F. (2003). Why visual attention and awareness are different. *Trends in Cognitive Sciences, 7*(1), 12–18.

Lamme, V. A. F. (2004). Separate neural definitions of visual consciousness and visual attention; a case for phenomenal awareness. *Neural Networks, 17*, 861–872.

Lamme, V. A. F. (2006a). Towards a true neural stance on consciousness. *Trends in Cognitive Sciences, 10*(11), 494–501.

Lamme, V. A. F. (2006b). Zap! Magnetic tricks on conscious and unconscious vision. *Trends in Cognitive Sciences, 10*(5), 193–195.

Landis, B. N., Frasnelli, J., Reden, J., Lacroix, J. S., & Hummel, T. (2005). Differences between orthonasal and retronasal olfactory functions in patients with loss of the sense of smell. *Archives of Otolaryngology—Head and Neck Surgery, 131*, 977–981.

Larsson, M., Oberg-Blavarg, C., & Jonsson, F. U. (2009). Bad odors stick better than good ones. *Experimental Psychology, 56*(6), 375–380. https://doi.org/10.1027/1618-3169.56.6.375

Laurent, R., & Chaix, R. (2012). MHC-dependent mate choice in humans: Why genomic patterns from the HapMap European American dataset support the hypothesis. *Bioessays, 34*, 267–271. https://doi.org/10.1002/bies.201100150

Lehrner, J. P., Glück, J., & Laska, M. (1999). Odor identification, consistency of label use, olfactory threshold and their relationship to odor memory over the human lifespan. *Chemical Senses, 24*, 337–346. https://doi.org/10.1093/chemse/24.3.337

Lenochova, P., Vohnoutova, P., Roberts, S. C., Oberzaucher, E., Grammer, K., and Havlicek, J. (2012). Psychology of fragrance use: Perception of individual odor and perfume blends reveals a mechanism for idiosyncratic effects on fragrance choice. *PLoS ONE, 7*, e33810. https://doi.org/10.1371/journal.pone.0033810

Le Norcy, S. (1991). Selling perfume: A technique or an art? In S. Van Toller & G. H. Dodd (Eds.), *Perfumery: The psychology and biology of fragrance* (pp. 217–226). Chapman and Hall.

Leopold, D. (2002). Distortions of olfactory perception. *Chemical Senses, 27*, 611–615.

Li, W., Howard, J. D., Parrish, T. B., & Gottfried, J. A. (2008). Aversive learning enhances perceptual and cortical discrimination of indiscriminable odor cues. *Science, 319*, 1842–1845.

Li, W., Moallem, I., Paller, K. A., & Gottfried, J. A. (2007). Subliminal smells can guide social preferences. *Psychological Science, 18*(12), 1044–1050.

Licon, C. C., Bosc, G., Sabri, M., Mantel, M., Fournel, A., Bushdid, C., Golebiowski, J., Robardet, C., Plantevit, M., Kaytoue, M., & Bensafi, M. (2019). Chemical features mining provides new descriptive structure-odor relationships. *PLoS Computational Biology, 15*(4), e1006945. https://doi.org/10.1371/journal.pcbi.1006945

Linster, C., & Cleland, T. A. (2013). Spatiotemporal coding in the olfactory system. *Computational Neuroscience, 9*, 229–242. https://doi.org/10.1007/978-1-4614-1424-7_11

Livermore, A., & Laing, D. G. (1996). Influence of training and experience on the perception of multicomponent odor mixtures. *Journal of Experimental Psychology: Human Perception and Performance, 22*, 267–277.

Livermore, A., & Laing, D. G. (1998). The influence of odor type on the discrimination and identification of odorants in multicomponent odor mixtures. *Physiology and Behavior, 65*(2), 311–320.

Lycan, W. G. (2000). The slighting of smell. In N. Bhushan & S. Rosenfeld (Eds.), *Of minds and molecules: New philosophical perspectives on chemistry.* (pp. 273–289) Oxford University Press.

Lycan, W. G. (2014). The intentionality of smell. *Frontiers in Psychology, 5*, 436.

Lycan, W. G. (2018). What does taste represent? *Australasian Journal of Philosophy, 96*(1), 28–37. https://doi.org/10.1080/00048402.2017.1291697

Lyman, B. J., & McDaniels, M. A. (1990). Memory for odors and odor names: Modalities of elaboration and imagery. *Journal of Experimental Psychology Learning Memory and Cognition, 16*(4), 656–664. https://doi.org/10.1037/0278-7393.16.4.656

Mackay-Sim, A., & Kittel, P. (1991). Cell dynamics in the adult mouse olfactory epithelium: A quantitative autoradiographic study. *Journal of Neuroscience, 11*(4), 979–984. https://doi.org/10.1523/JNEUROSCI.11-04-00979.1991

Mackay-Sim, A., & Royet, J. (2006). The structure and function of the olfactory system. In W. J. Brewer, D. Castle, & C. Pantelis (Eds.), *Olfaction and the brain.* (pp. 3–27) Cambridge University Press.

Macpherson, F. (2011a). Individuating the senses. In F. Macpherson (Ed.), *The senses.* (pp. 3–46) Oxford University Press.

Macpherson, F. (2011b). Cross-modal experiences. *Proceedings of the Aristotelian Society, 111*, 429–468. https://doi.org/10.1111/j.1467-9264.2011.00317.x

Mainland, J. D., Keller, A., Li, Y. R., Zhou, T., Trimmer, C., Snyder, L. L., Moberly, A. H., Adipietro, K. A., Liu, W. L. L., Zhuang, H., Zhan, S., Lee, S. S., Lin, A., & Matsunami, H. (2014). The missense of smell: Functional variability in the human odorant receptor repertoire. *Nature Neuroscience, 17*, 114–120.

Mainland, J., & Sobel, N. (2006). The sniff is part of the olfactory percept. *Chemical Senses, 31*, 181–196.

Majid, A. (2015). Cultural factors shape olfactory language. *Trends in Cognitive Sciences, 19*, 629–630. https://doi.org/10.1016/j.tics.2015.06.009

Majid, A. (2021). Human olfaction at the intersection of language, culture, and biology. *Trends in Cognitive Sciences, 25*(2), 111–123. https://doi.org/10.1016/j.tics.2020.11.005

Majid, A., & Burenhult, N. (2014). Odors are expressible in language, as long as you speak the right language. *Cognition, 130*, 266–270. https://doi.org/10.1016/j.cognition.2013.11.004

Majid, A., & Kruspe, N. (2018). Hunter-gatherer olfaction is special. *Current Biology, 28*, 409–413.

Majid, A., Roberts, S. G., Cilissen, L., & Levinson, S. C. (2018). Differential coding of perception in the world's languages. *Proceedings of the National Academy Sciences of the United States of America, 115*(45), 11369–11376. https://doi.org/10.1073/pnas.1720419115

Malnic, B., Hirono, J., Sato, T., & Buck, L. B. (1999). Combinatorial receptor codes for odors. *Cell, 96*, 713–723.

Mandik, P. (2000). *Objective subjective—Allocentric and egocentric representations in thought and experience* [Unpublished doctoral dissertation]. Washington University.

Mandik, P. (2005). Phenomenal consciousness and the allocentric–egocentric interface. In R. Buccheri, A. C. Elitzur, & M. Saniga (Eds.), *Endophysics, time, quantum and the subjective*, Proceedings of the ZiF Interdisciplinary Research Workshop, World Scientific Publishing Co. pp. 463–485.

Mandik, P. (2009). The neurophilosophy of subjectivity. In J. Bickle (Ed.), *Oxford handbook of philosophy and neuroscience*.(pp. 601–618) Oxford University Press.

Mantel, M., Ferdenzi, C., Roy, J., & Bensafi, M. (2019). Individual differences as a key factor to uncover the neural underpinnings of hedonic and social functions of human olfaction—Current findings from PET and fMRI studies and future considerations. *Brain Topography, 32*(6), 977–986. https://doi.org/10.1007/s10548-019-00733-9

Mantel M, Roy JM, Bensafi M. (2021). Accounting for Subjectivity in Experimental Research on Human Olfaction. *Chem Senses*. 1;46:bjaa082. doi: 10.1093/chemse/bjaa082.

Margolis, E., & Laurence, S. (1999). *Concepts*. MIT Press.

Marr, D. (1982). *Vision*. Freeman.

Marshall, W. H., Woolsey, C. N., & Bard, P. (1941). Observations on cortical somatic sensory mechanisms of cat and monkey. *Journal of Neurophysiology, 4*, 1–24.

Martinez, M. (2015). Disgusting smells and imperativism. *Journal of Consciousness Studies, 22*(5–6), 191–200.

Martinez-Manrique, F. (2014). Systematicity and conceptual pluralism. In P. Calvo & J. Symons (Eds.), *The architecture of cognition: Rethinking Fodor and Pylyshyn's systematicity challenge* (pp. 305–334). MIT Press.

Mashour GA, Roelfsema P, Changeux JP, Dehaene S. Conscious Processing and the Global Neuronal Workspace Hypothesis. Neuron. 2020 Mar 4;105(5):776–798. doi: 10.1016/j.neuron.2020.01.026.

Matthen, M. (2015). Individuation of the senses. In M. Matthen (Ed.), *Oxford handbook of philosophy of perception* (pp. 567–586). Oxford University Press.

Matthen, M. (2017). Is perceptual experience normally multimodal? In B. Nanay (Ed.), *Current controversies in philosophy of perception*. Routledge.

Maye, A. & Engel, A. K. (2012). Neuronal assembly models of compostionality. In M. Werning, W. Hinzen & E. Machery (Eds.), The Oxford handbook of compositionality. (pp. 616–632) Oxford: Oxford University Press.

Mayhew, E. J., Arayata, C. J., Gerkin, R. C., Lee, B. K., Magill, J. M., Snyder, L. L., Little, K. A., Yu, C. W., & Mainland, J. D. (2022). Transport features predict if a molecule is odorous. *Proceedings of the National Academy of Sciences of the United States of America, 119*(15), e2116576119. https://doi.org/10.1073/pnas.2116576119

McAlonan, K., Brown, V. J., & Bowman, E. M. (2000). Thalamic reticular nucleus activation reflects attentional gating during classical conditioning. *Journal of Neuroscience, 20*, 8897–8901.

McBride, S. A., & Slotnick, B. (1997). The olfactory thalamocortical system and odor reversal learning examined using an asymmetrical lesion paradigm in rats. *Behavioral Neuroscience,* 111(6), 1273–1284.

McFarland, D. J., & Sibly, R. M. (1975). The behavioural final common path. *Philosophical Transactions of the Royal Society of London, Series B Biological Sciences, 270*(907), 265–293.

McNamara, A. M., Magidson, P. D., & Linster, C. (2007). Binary mixture perception is affected by concentration of odor components. *Behavioral Neuroscience, 121*(5), 1132–1136.

McPeek, R. M., & Keller, E. L. (2002). Saccade target selection in the superior colliculus during a visual search task. *Journal of Neurophysiology, 88*(4), 2019–2034.

Meierhenrich, U. J., Golebiowski, J., Fernandez, X., & Cabrol-Bass, D. (2004). The molecular basis of olfactory chemoreception. *Angewandte Chemie, 43*, 6410–6412.

Meister, M. (2015). On the dimensionality of odor space. *eLife, 4*, e07865. https://doi.org/10.7554/eLife.07865

Melcher, J. M., & Schooler, J. W. (1996). The misremembrance of wines past: Verbal and perceptual expertise differentially mediate verbal over-shadowing of taste memory. *Journal of Memory and Language, 35*, 231–245. https://doi.org/10.1006/jmla.1996.0013

Meredith, M. (1991). Sensory processing in the main and accessory olfactory systems: Comparisons and contrasts. *Journal of Steroid Biochemistry and Molecular Biology, 39*, 601–614.

Merker, B. (2007). Consciousness without a cerebral cortex: A challenge for neuroscience and medicine. *Behavioral and Brain Sciences, 30*(1), 63–134.

Michel, M., Beck, D., Block, N., Blumenfeld, H., Brown, R., Carmel, D., Carrasco, M., Chirimuuta, M., Chun, M., Cleeremans, A., Dehaene, S., Fleming, S. M., Frith, C., Haggard, P., He, B. J., Heyes, C., Goodale, M. A., Irvine, L., Kawato, M., . . . Yoshida, M. (2019). Opportunities and challenges for a maturing science of consciousness. *Nature Human Behaviour, 3*, 104–107.

Milinski, M., Croy, I., Hummel, T., & Boehm, T. (2013). Major histocompatibility complex peptide ligands as olfactory cues in human body odour assessment. *Proceedings of the Royal Society B: Biological Sciences, 280*, 20122889. https://doi.org/10.1098/rspb.2012.2889

Milinski, M., & Wedekind, C. (2001). Evidence for MHC-correlated perfume preferences in humans. *Behavioral Ecology, 12*, 140–149. https://doi.org/10.1093/beheco/12.2.140

Millar, B. (2017). Smelling objects. *Synthese, 196*, 4279–4303. https://doi.org/10.1007/s11229-017-1657-8

Millikan, R. G. (1984). *Language, thought, and other biological categories*. MIT Press.

Millikan, R. G. (1993). *White Queen psychology and other essays for Alice*. MIT Press.

Millikan, R. G. (1994). Biosemantics. In S. P. Stich & T. A. Warfield (Eds.), *Mental representation*. Blackwell Publishers.

Millikan, R. G. (2000). *On clear and confused ideas*. Cambridge University Press.

Mion, M., Patterson, K., Acosta-Cabronero, J., Pengas, G., Izquierdo-Garcia, D., Hong, Y., Fryer, T. D., Williams, G. B., Hodges, J. R., & Nestor, P. J. (2010). What the left and right anterior fusiform gyri tell us about semantic memory. *Brain, 133*, 3256–3268.

Mirza, N., Kroger, H., & Doty, R. (1997). Influence of age on the "nasal cycle." *Laryngoscope, 107*, 62–66.

Miwa, T., Furukawa, M., Tsukatani, T., Costanzo, R. M., Di-Nardo, L. J., & Reiter, E. R. (2001). Impact of olfactory impairment on quality of life and disability. *Archives of Otolaryngology—Head and Neck Surgery, 127*(5), 497–503.

Møller, P., Wulff, C., & Köster, E. P. (2004). Do age differences in odour memory depend on differences in verbal memory? *Neuroreport, 15*, 915–917. https://doi.org/10.1097/00001756-200404090-00036

Morrison, E. E., & Costanzo, R. M. (1990). Morphology of the human olfactory epithelium. *Journal of Comparative Neurology, 297*, 1–13.

Morrison, E. E., & Costanzo, R. M. (1992). Morphology of olfactory epithelium in humans and other vertebrates. *Microscopy Research and Technique, 23*(1), 49–61.

Mort, E., Cairns, S., Hersch, H., & Finlay, B. (1980). The role of the superior colliculus in visually guided locomotion and visual orienting in the hamster. *Physiological Psychology, 8*, 20–28.

Moss, H. E., Rodd, J. M., Stamatakis, E. A., Bright, P., & Tyler, L. K. (2005). Anteromedial temporal cortex supports fine-grained differentiation among objects. *Cerebral Cortex, 15*, 616–627.

Mozell, M., Kent, P., & Murphy, S. (1991). The effect of flow rate upon the magnitude of the olfactory response differs for different odorants. *Chemical Senses, 16*, 631–649.

Murakami, M., Kashiwadani, H., Kirino, Y., & Mori, K. (2005). State-dependent sensory gating in olfactory cortex. *Neuron, 46*, 285–296.

Murphy, C., Cain, W. S., Gilmore, M. M., & Skinner, R. B. (1991). Sensory and semantic factors in recognition memory for odors and graphic stimuli: Elderly versus young persons. *American Journal of Psychology, 104*(2), 161–192.

Murphy, G. L. (2002). *The big book of concepts*. MIT Press.

Mylopolous, M. (2022). Neurobiological theories of consciousness. In B. D. Young & C. Dicey-Jennings (Eds.), *Mind, cognition, and neuroscience*. (pp. 280–293) Routledge.

Nagel, T. (1974). What is it like to be a bat? *Philosophical Review, 83*(4), 435–450.

Nordlander, C., Hammarstrom, L., Lindblom, B., & Edvard Smith, C. I. E. (1983). No role of HLA in mate selection. *Immunogenetics, 18*, 429–431.

Ober, C., Weitkamp, L. R., Cox, N., Dytch, H., Kostyu, D., & Elias, S. (1997). HLA and mate choice in humans. *American Journal of Human Genetics, 61*, 497–504.

O'Callaghan, C. (2007). *Sounds*. Oxford University Press.

O'Callaghan, C. (2008). Object perception: Vision and audition. *Philosophy Compass, 3*(4), 803–829.

O'Callaghan, C. (2015). The multisensory character of perception. *Journal of Philosophy, 112*, 551–569.

O'Callaghan, C. (2016). Objects for multisensory perception. *Philosophical Studies, 173*, 1269–1289.

Odegaard, B., Knight, R. T., & Lau, H. (2017). Should a few null findings falsify prefrontal theories of conscious perception? *Journal of Neuroscience, 37*(40), 9593–9602. https://doi.org/10.1523/JNEUROSCI.3217-16.2017

Ohloff, G. (1986). Chemistry of odor stimuli. *Experientia, 42*, 271–279.

Oizumi, M., Albantakis, L., & Tononi, G (2014). From the phenomenology to the mechanisms of consciousness: Integrated information theory 3.0. *PLoS Computational Biology, 10*(5), e1003588. https://doi.org/10.1371/journal.pcbi.1003588

Okamoto, K., Shiga, H., Nakamura, H., Matsui, M., & Miwa, T. (2020). Relationship between olfactory disturbance after acute ischemic stroke and latent thalamic hypoperfusion. *Chemical Senses, 45*(2), 111–118. https://doi.org/10.1093/chemse/bjz077

Olofsson, J. K. (2014). Time to smell: A cascade model of human olfactory perception based on response-time (RT) measurement. *Frontiers in Psychology, 5*, 33.

Olofsson, J. K., Bowman, N. E., & Gottfried, J. A. (2013). High and low roads to odor valence? A choice response-time study. *Journal of Experimental Psychology: Human Perception and Performance, 39*, 1205–1211.

Olofsson, J. K., Bowman, N. E., Khatibi, K., & Gottfried, J. A. (2012). A time-based account of the perception of odor objects and valences. *Psychological Science, 23*, 1224–1232.

Olofsson, J. K., Ekström, I., Larsson, M., & Nordin, S. (2021). Olfaction and aging—A review of the current state of research and future directions. *I-Perception, 12*(3), 1–24. https://doi.org/10.1177/20416695211020331

Olofsson, J. K., & Gottfried, J. A. (2015a). The muted sense: Neurocognitive limitations of olfactory language. *Trends in Cognitive Science, 19*, 314–321.

Olofsson, J. K., & Gottfried, J. A. (2015b). Response to Majid: Neurocognitive and cultural approaches to odor naming are complementary. *Trends in Cognitive Science, 19*, 630–631.

Olofsson, J. K., Hurley, R. S., Bowman, N. E., Bao, X., Mesulam, M. M., & Gottfried, J. A. (2014). A designated odor-language integration system in the human brain. *Journal of Neuroscience, 34*, 14864–14873. https://doi.org/10.1523/JNEUROSCI.2247-14.2014

Olofsson, J. K. Rogalski, E., Harrison, T., Mesulam, M. M., & Gottfried, J. A. (2013). A cortical pathway to olfactory naming: Evidence from primary progressive aphasia. *Brain, 136*, 1245–1259.

Olofsson, J. K., & Wilson, D. A. (2018). Human olfaction: It takes two villages. *Current Biology, 28*, R103–R126.

O'Meara, C., & Majid, A. (2016). How changing lifestyles impact Seri smellscapes and smell language. *Anthropological Linguistics, 58*, 107–131. https://doi.org/10.1353/anl.2016.0024

O'Meara, C., & Majid, A. (2020). Anger stinks in Seri: Olfactory metaphor in a lesser-described language. *Cognitive Linguistics, 31*(3), 367–391. https://doi.org/10.1515/cog-2017-0100

Ongur, D., & Price, J. L. (2000). The organization of networks within the orbital and medial prefrontal cortex of rats, monkeys and humans. *Cerebral Cortex, 10*(3), 206–219.

Pagin, P., & Westerståhl, D. (2010a). Compositionality I: Definitions and variants. *Philosophy Compass, 5*, 265–282.

Pagin, P., & Westerståhl, D. (2010b). Compositionality II: Arguments and problems. *Philosophy \Compass, 5*, 250–264.

Papineau, D. (1987). *Reality and representation*. Blackwell.

Papineau, D. (1993). *Philosophical naturalism*. Blackwell.

Parr, W. V., Heatherbell, D., & White, K. G. (2002). Demystifying wine expertise: Olfactory threshold, perceptual skill and semantic memory in expert and novice wine judges. *Chemical Senses, 27*, 747–755. https://doi.org/10.1093/chemse/27.8.747

Patterson, K., Nestor, P. J., & Rogers, T. T. (2007). Where do you know what you know? The representation of semantic knowledge in the human brain. *Nature Reviews Neuroscience, 8*, 976–988.

Pause, B. M., Sojka, B., & Ferstl, R. (1997). Central processing of odor concentration is a temporal phenomena as revealed by chemosensory event-related potentials. *Chemical Senses, 22*, 9–26.

Pautz, A. (2021). *Perception*. Routledge.

Pelosi, P. (2001). The role of perireceptor events in vertebrate olfaction. *Cellular and Molecular Life Sciences, 58*(4), 503–509.

Pevsner, J., Trifiletti, R. R., Strittmatter, S. M., Sklar, P. B., & Snyder, S. H. (1985). Purification and characterization of a pyrazine odorant binding protein. *Chemical Senses, 10*, 397.

Phillips, I. & Block, N. (2016). Debate on unconscious perception. In Bence Nanay (ed.), Current Controversies in Philosophy of Perception. (pp. 165–192) Routledge.

Piantadosi, S., Tenenbaum, J. B., & Goodman, N. D. (2016). The logical primitives of thought: Empirical foundations for compositional cognitive models. *Psychological Review, 123*, 392–424.

Plailly, J., Howard, J. D., Gitelman, D. R., & Gottfried, J. A. (2008). Attention to odor modulates thalamocortical connectivity in the human brain. *Journal of Neuroscience, 28*(20), 5257–5267.

Pollack, M. S., Wysocki, C. J., Beauchamp, G. K., Braun, D., Callaway, C., & Dupont, B. (1982). Absence of HLA association or linkage for variations in sensitivity to the odor of androstenone. *Immunogenetics, 15*, 579–589.

Porter, J., Anand, T., Johnson, B. N., Khan, R. M., & Sobel, N. (2005). Brain mechanisms for extracting spatial information from smell. *Neuron, 47*, 581–592.

Porter, J., Craven, B., Khan, R. M., Chang, S. J., Kang, I., Judkewitz, B., Volpe, J., Settles, G., & Sobel, N. (2007). Mechanisms of scent-tracking in humans. *Nature Neuroscience, 10*(1), 27–29.

Porter, R. H., Balog, R. D., Cernoch, J. M., & Franchi, C. (1986). Recognition of kin through characteristic body odors. *Chemical Senses, 11*, 389–395.

Porter, R. H., Cernoch, J. M., & McLaughlin, F. J. (1983). Maternal recognition of neonates through olfactory cues. *Physiology and Behavior, 30*(1), 151–154.

Porter R. H., & Winberg J. (1999). Unique salience of maternal breast odors for newborn infants. *Neuroscience and Biobehavioral Reviews, 23*, 439–449.

Prescott, J. (2015). Multisensory processes in flavour perception and their influence on food choice. *Current Opinion in Food Science, 3*, 47–52.

Price, A. R., Bonner, M. F., Peelle, J. E., & Grossman, M. (2015a). Converging evidence for the neuroanatomic basis of combinatorial semantics in the angular gyrus. *Journal of Neuroscience, 35*, 3276–3284.

Price, A. R., McAdams, H., Grossman, M., & Hamilton, R. H. (2015b). A meta-analysis of transcranial direct current stimulation studies examining the reliability of effects on language measures. *Brain Stimulation, 8*, 1093–1100.

Price, A. R., Peelle, J. E., Bonner, M. F., Grossman, M., & Hamilton, R. H. (2016). Causal evidence for a mechanism of semantic integration in the angular gyrus as revealed by high-definition transcranial direct current stimulation. *Journal of Neuroscience, 6*, 3829–3838.

Price, J. L. (1973). An autoradiographic study of complementary laminar patterns of termination of afferent fibers to the olfactory cortex. *Journal of Comparative Neurology, 150*, 87–108.

Price, J. L. (1987). The central and accessory olfactory. In T. E. Finger & W. L. Silver (Eds.), *Neurobiology of taste and smell* (pp. 179–203). Wiley.

Price, J. L. (1990). Olfactory system. In G. Paxinos (Ed.), *The human nervous system* (pp. 979–1001). Academic Press.

Price, J. L., & Slotnick, B. M. (1983). Dual olfactory representation in the rat thalamus: An anatomical and electrophysiological study. *Journal of Comparative Neurology, 215*, 63–77.

Price, J. L., Slotnick, B. M., & Revial, M. F. (1991b). Olfactory projections to the hypothalamus. *Journal of Comparative Neurology, 306*, 447–461.

Principato, J. J., & Ozenberger, J. M. (1970). Cyclical changes in nasal resistance. *Archives of Otolaryngology—Head and Neck Surgery, 91*, 71–77.

Prinz, J. (2000). A Neurofunctional Theory of Visual Consciousness, *Consciousness and Cognition*, 9, (2), pp. 243–259. https://doi.org/10.1006/ccog.2000.0442.

Prinz, J. (2002). *Furnishing the mind.* MIT Press.

Prinz, J. (2005). A neurofunctional theory of consciousness. In A. Brooks (Ed.), *Philosophy and neuroscience.* (pp. 381–396) Cambridge University Press.

Prinz, J. (2007). The intermediate-level theory of consciousness. In M. Velmans & S. Schneider (Eds.), *The Blackwell companion to consciousness.* (pp. 248—260) Blackwell.

Prinz, J. (2012). Regaining composure: A defense of prototype compositionality. In M. Werning, W. Hinzen & E. Machery (Eds.), *The Oxford handbook of compositionality.* (pp. 437–454) Oxford University Press.

Pylyshyn, Z. (2003). Return of the mental image: Are there really pictures in the brain? *Trends in Cognitive Sciences, 7*(3), 113–118.

Qu, L. P., Kahnt, T., Cole, S. M., & Gottfried, J. A. (2016). De novo emergence of odor category representations in the human brain. *Journal of Neuroscience, 36*, 468–478.

Quilty-Dunn, J. (2013). Reid on olfaction and secondary qualities. *Frontiers in Psychology, 4*, 974. https://doi.org/10.3389/fpsyg.2013.00974

Rabin, M. D. (1988). Experience facilitates olfactory quality discrimination. *Perception and Psychophysics, 44*, 532–540. https://doi.org/10.3758/BF03207487

Raithel, C. U., & Gottfried, J. A. (2021). Using your nose to find your way: Ethological comparisons between human and non-human species. *Neuroscience and Biobehavioral Reviews, 128*, 766–779. https://doi.org/10.1016/j.neubiorev.2021.06.040

Rajagopalan, A., & Assisi, C. (2020). Effect of circuit structure on odor representation in the insect olfactory system. *eNeuro, 7*(3), ENEURO.0130-19.2020. https://doi.org/10.1523/ENEURO.0130-19.2020

Raverby, I., Snitz, K., & Sobel, N. (2022). There is chemistry in social chemistry. *Science Advances, 8*, eabn0154.

Reid, T. (1997). *An inquiry into the human mind*. D. R. Brookes (Ed.). Edinburgh University Press. (Original work published 1764)

Rennaker, R. L., Chen, C. F., Ruyle, A. M., Sloan, A. M., & Wilson, D. A. (2007). Spatial and temporal distribution of odorant-evoked activity in the piriform cortex. *Journal of Neuroscience, 27*, 1534–1542.

Rice, C. (2013). Concept empiricism, content, and compositionality. *Philosophical Studies, 162*, 567–583. https://doi.org/10.1007/s11098-011-9782-6

Richardson, L. (2013a). Flavor, taste, and smell. *Mind and Language, 28*, 322–341.

Richardson, L. (2013b). Sniffing and smelling. *Philosophical Studies, 162*, 401–419.

Rinck, F., Rouby, C., & Besafi, M. (2009). Which format for odor images? *Chemical Senses, 34*, 11–13.

Robert-Hazotte, A., Faure, P., Neiers, F., Potin, C., Artur, Y., Coureaud, G., & Heydel, J. M. (2019b). Nasal mucus glutathione transferase activity and impact on olfactory perception and neonatal behavior. *Scientific Reports, 9*, 3104. https://doi.org/10.1038/s41598-019-39495-6

Robert-Hazotte, A., Schoumacker, R., Semon, E., Briand, L., Guichard, E., Le Quéré, J. L., Faure, P., & Heydel, J. M. (2019a). Ex vivo real-time monitoring of volatile metabolites resulting from nasal odorant metabolism. *Scientific Reports, 9*, 2492.

Roberts, S. C., Havlíček, J., & Schaal, B. (2020). Human olfactory communication: Current challenges and future prospects. *Philosophical Transactions of the Royal Society B, 375*, 20190258. https://doi.org/10.1098/rstb.2019.0258

Romagny, S., Thomas-Danguin, T., & Coureaud, G. (2015). Configural processing of odor mixture: Does the learning of elements prevent the perception of configuration in the newborn rabbit? *Physiology and Behavior, 142*, 161–169. https://doi.org/10.1016/j.physbeh.2015.02.019

Rosenberg, L. T., Cooperman, D., & Payne, R. (1983). HLA and mate selection. *Immunogenetics, 17*, 89–93. https://doi.org/10.1007/BF00364292

Rosenthal, D. M. (2002). How many kinds of consciousness? *Consciousness and Cognition, 11*, 653–665.

Rosenthal, D. M. (2007). *Consciousness and mind.* Oxford University Press.

Rosenthal, D. M. (2009). Concepts and definition of consciousness. In W. P. Banks (Ed.), *Encyclopedia of consciousness.* **(pp. 157–169)** Elsevier.

Rosenthal, D. M. (2010). How to think about mental qualities. *Philosophical Issues: Philosophy of Mind, XX*, 368–393.

Rossiter, K. J. (1996). Structure–odor relationships. *Chemical Reviews, 96*, 3201–3240.

Rouby, C., Bourgeat, F., Rinck, F., Poncelet, J., & Bensafi, M. (2009). Perceptual and sensorimotor differences between "Good" and "Poor" olfactory mental imagers. international symposium on olfaction and taste. *Annals of the New York Academy of Sciences, 1170*, 333–337. https://doi.org/10.1111/j.1749-6632.2009.03915.x

Rouquier, S., Blancher, A., & Giorgi, D. (2000). The olfactory receptor gene repertoire in primates and mouse: Evidence for reduction of the functional fraction in primates. *Proceedings of the National Academy of Sciences of the United States of America, 97*(6), 2870–2874. https://doi.org/10.1073/pnas.040580197

Rozin, E. (1983) Ethnic cuisine: the *Flavor- principle cookbook.* S. Greene Press.

Rozin, P. (1978). The use of characteristic flavourings in human culinary practise. In C. M. Apt (Ed.), *Flavour: Its chemical behavioural and commercial aspects* (pp. 101–127). Westview Press.

Rozin, P. (1982). Taste-smell confusions and the duality of the olfactory sense. *Perception and Psychophysics, 31*, 397–401.

Rozin, P., Hammer, L., Oster, H., & Marmora, V. (1986). The child's conception of food. *Appetite, 7*, 141–151.

Russell, M. J. (1976). Human olfactory communication. *Nature, 260*, 520–522.

Sans, M., Alvarez, I., Callegari-Jacques, S. M., & Salzano, F. M. (1994). Genetic similarity and mate selection in Uruguay. *Journal of Biosocial Science, 26*, 285–289.

Sapolsky, R. M., & Eichbaum, H. (1980). Thalamocortical mechanisms in odor-guided behavior. II. Effects of lesions of the medio-dorsal thalamic nucleus and frontal cortex on odor preferences and sexual behavior in the hamster. *Brain Behavior and Evolution, 17*(4), 276–290.

Savic, I., & Berguland, H. (2004). Passive perception of odors and semantic circuits. *Human Brain Mapping, 21*, 271–278.

Savic, I., Gulyas, B., Larsson, M., & Roland, P. (2000). Olfactory functions are mediated by parallel and hierarchical processing. *Neuron, 26*, 735–745.

Schaal, B. (2012). Emerging chemosensory preferences. In G. M. Zucco, R. S. Herz, & B. Schaal (Eds.), *Olfactory cognition*. (pp. 137–268). John Benjamin.

Schaal, B., Saxton, T. K., Loos, H., Soussignan, R., & Durand, K. (2020). Olfaction scaffolds the developing human from neonate to adolescent and beyond. *Philosophical Transactions of the Royal Society B, 375*, 20190261. https://doi.org/10.1098/rstb.2019.0261

Schiller, P. H., & Lee, K. (1994). The effects of lateral geniculate nucleus, area V4, and middle temporal (MT) lesions on visually guided eye movements. *Visual Neuroscience, 11*, 229–241.

Schiller, P. H., True, S. D., & Conway, J. L. (1979). Effects of frontal eye field and superior colliculus ablations on eye movements. *Science, 206*, 590–592.

Schneider, G. E. (1967). Contrasting visuomotor functions of tectum and cortex in the golden hamster. *Psychologische Forschung, 31*, 52–62.

Schopenhauer, A. (1891). *Studies in pessimism*. Swan Sonnenschein & Co.

Schwartz, G. E. R. (2000). Individual differences in subtle awareness a levels of awareness. In R. G. Kunzendorf & B. Wallace (Eds.), *Individual differences in conscious experience, advances in consciousness research*. (pp. 209–225). John Benjamins B.V.

Schwartz, G. E. R., Bell, I. R., Dikman, Z. V., Fernandez, M., Kline, J. P., Peterson, J. M., & Wright, K. P. (1994). EEG responses to low-level chemicals in normals and cacosmics. *Toxicology and Industrial Health, 10*, 633–643.

Schwitzgebel, E. (2008). The unreliability of naive introspection. *Philosophical Review, 117*(2), 245–273.

Schwitzgebel, E. (2012). Introspection, what? In D. Smithies & D. Stoljar (Eds.), *Introspection and consciousness*. Oxford University Press.

Sela, L., Sacher, Y., Serfaty, C., Yeshurun, Y., Soroker, N., & Sobel, N. (2009). Spared and impaired olfactory abilities after thalamic lesions. *Journal of Neuroscience, 30*, 12059–12069.

Sela, L., & Sobel, N. (2010). Human olfaction: A constant state of change-blindness. *Experimental Brain Research, 205*(1), 13–29.

Shepherd, G. M. (1972). Synaptic organization of the mammalian olfactory bulb. *Physiological Reviews, 52*, 864–869.

Shepherd, G. M. (2012). *Neurogastronomy*. Columbia University Press.

Shipley, M. T. (1995). Olfactory system. In G. Paxinos (Ed.), *The rat nervous system* (2nd ed., pp. 899–928). Academic Press.

Siegel, S. (2021). The contents of perception. In E. N. Zalta (Ed.), *The Stanford encyclopedia of philosophy* (Fall 2021 Edition). https://plato.stanford.edu/archives/fall2021/entries/perception-contents/

Sinding, C., Thomas-Danguin, T., Chambault, A., Béno, N., Dosne, T., Chabanet, C., Schaal, B., & Coureaud, G. (2013). Rabbit neonates and human adults perceive a blending 6- component odor mixture in a comparable manner. *PLoS One, 8*(1), e53534.

Sjölund, S., Larsson, M., Olofsson, J. K., Suebert, J., & Laukka, E. (2017). Phantom smells: Prevalence and correlates in a population-based sample of older adults. *Chemical Senses, 42*, 309–318.

Skrzypulec, B. (2018). Olfactory objecthood. *Philosophia, 47*(3), 881–900. https://doi.org/10.1007/s11406-018-0017-3

Skrzypulec, B. (2019). The nonclassical mereology of olfactory experiences. *Synthese, 198*, 867–886. https://doi.org/10.1007/s11229-018-02072-x

Skrzypulec, B. (2021). Constitutivity in flavour perception. *Erkenntnis*. https://doi.org/10.1007/s10670-021-00503-9

Skrzypulec, B. (2022). Tracking representationalism and olfaction. *Mind and Language, 38*(2), 446–463.

Skrzypulec, B. (2023). Seeing and hearing flavours. In A. Keller & B. D. Young (Eds.), *Theoretical perspectives on smell*. (pp. 241–259). Routledge.

Slotnick, B. M., & Kaneko, N. (1981). Role of mediodorsal thalamic nucleus in olfactory discrimination learning in rats. *Science, 214*, 91–92.

Slotnick, B. M., & Schoonover, F. W. (1992). Olfactory pathways and the sense of smell. *Neuroscience and Biobehavioral Reviews, 16*, 453–472.

Small, D. M. (2012). Flavor is in the brain. *Physiology and Behavior, 107*, 540–552.

Smith, A. D. (2002). *The problem of perception*. Harvard University Press.

Smith, B. C. (2008). Same compounds: Different flavors. In D. Chassagne (Ed.), *Proceedings of Wine Active Compounds* (pp. 98–102). Oenopluria Media.

Smith, B. C. (2009). The objectivity of tastes and tasting. In B. Smith (Ed.), *Questions of taste*. (pp. 41–61). Oxford University Press.

Smith, B. C. (2013). Taste philosophical perspectives. In H. Pashler (Ed.), *The encyclopaedia of mind*. (pp. 731–735). Sage.

Smith, B. C. (2015). The chemical senses. In M. Matthen (Ed.), *Oxford handbook of philosophy of perception* (pp. 567–586). Oxford University Press.

Smith, B. C. (2023). The role of smell in consciousness. In A. Keller & B. D. Young (Eds.), *Theoretical perspective on smell.* (pp. 13–35). Routledge.

Smith, B. C., Sester, C., Ballester, J., & Deroy, O. (2017). The perceptual categorization of blended and single malt Scotch whiskies. *Flavour, 6,* 5. https://doi.org/10.1186/s13411-017-0056-x

Smythies, J. (1997). The functional neuroanatomy of awareness: With a focus on the role of various anatomical systems in the control of intermodal attention. *Consciousness and Cognition, 6,* 455–481.

Snitz, K., Yablonka, A., Weiss, T., Frumin, I., Khan, R. M., & Sobel, N. (2013). Predicting odor perceptual similarity from odor structure. *PLoS Computational Biology, 9*(9), e1003184.

Sobel, N., Johnson, B. N., Mainland, J., & Yousem, D. M. (2003). Functional neuroimaging of human olfaction. In R. L. Doty (Ed.), *Handbook of olfaction and gustation* (2nd ed., pp. 251–273). Marcel Dekker.

Sobel, N., Khan, R. M., Saltman, A., Sullivan, E. V., & Gabrieli, J. D. (1999a). The world smells different to each nostril. *Nature, 402,* 35.

Sobel, N., Prabhakaran, V., Hartley, C. A., Desmond, J. E., Glover, G. H., Sullivan, E. V., & Gabrieli, J. D. (1999b). Blind smell: Brain activation induced by an undetected air-borne chemical. *Brain, 122*(2), 209–217.

Solomon, G. E. A. (1990). Psychology of novice and expert wine talk. *American Journal of Psychology, 103*(4), 495–517.

Speed, L. J., & Majid, A. (2018). An exception to mental simulation no evidence for embodied odor language. *Cognitive Science, 42,* 1146–1178. https://doi.org/10.1111/cogs.12593

Speed, L. J., & Majid, A. (2020). Grounding language in the neglected senses of touch, taste, and smell. *Cognitive Neuropsychology, 37*(5–6), 363–392. https://doi.org/10.1080/02643294.2019.1623188

Spence, C. (2015). Multisensory flavor perception. *Cell, 161,* 24–35.

Spence, C., Auvray, M., & Smith, B. C. (2015). Confusing tastes with flavours. In D. Stokes, M. Matthen, & S. Biggs (Eds.), *Perception and its modalities* (pp. 247–274). Oxford University Press.

Spence, C., McGlone, F. P., Kettenmann, B., & Kobal, G. (2001). Attention to olfaction: A psychophysical investigation. *Experimental Brain Research, 138,* 432–437.

Stettler, D. D., & Axel, R. (2009). Representations of odor in the piriform cortex. *Neuron, 63,* 854–864.

Stevenson, R. J. (2009a). *The psychology of flavor.* Oxford University Press.

Stevenson, R. J. (2009b). An initial evaluation of the functions of human olfaction. *Chemical Senses, 35*(1), 3–20.

Stevenson, R. J. (2009c). Phenomenal and access consciousness in olfaction. *Consciousness and Cognition, 18*(4), 1–14.

Stevenson, R. J. (2014). Object concepts in the chemical senses. *Cognitive Science, 38*, 1360–1383. https://doi.org/10.1111/cogs.12111

Stevenson, R. J., & Case, T. I. (2005). Olfactory imagery: A review. *Psychonomic Bulletin and Review, 12*(2), 244–264.

Stevenson, R. J., & Mahmut, M. K. (2013). The accessibility of semantic knowledge for odours that can and cannot be named. *Quarterly Journal of Experimental Psychology, 66*, 1414–1431.

Stewart, T., & Eliasmith, C. (2012). Compositionality and biologically plausible models. In M. Werning, W. Hinzen, & E. Machery (Eds.), *The Oxford handbook of compositionality*. (pp. 596–615). Oxford University Press.

Storm, J. F., Boly, M., Casali, A. G., Massimini, M., Olcese, U., Pennartz, C. M. A., & Wilke, M. (2017). Consciousness regained—Disentangling mechanisms, brain systems, and behavioral responses. *Journal of Neuroscience, 37*(45), 10882–10893. https://doi.org/10.1523/JNEUROSCI.1838-17.2017

Sulmont-Rosse, C. (2005). Odor naming methodology: Correct identification with multiple-choice versus repeatable identification in a free task. *Chemical Senses, 30*, 23–27.

Tahirova, N., Poivet, E., Xu, L., Peterlin, Z., Zou, D. J., & Firestein, S. S. (2019). *Bioisosterism reveals new structure-odor relationships*. bioRxiv. https://doi.org/10.1101/567701

Talbot, S. A., & Marshall, W. H. (1941). Physiological studies on neural mechanisms of visual localization and discrimination. *American Journal of Ophthalmology, 24*, 1255–1263.

Taubes, G. (2016). *The case against sugar*. Knopf.

Teghtsoonian, M. (1982). Perceived effort in sniffing: The effects of sniff pressure and resistance. *Attention, Perception and Psychophysics, 31*(4), 324–329.

Teghtsoonian, R., & Teghtsoonian, M. (1984). Testing a perceptual constancy model for odor strength: The effects of sniff pressure and resistance to sniffing. *Perception, 13*, 743–752.

Teghtsoonian, R., Teghtsoonian, M., Berglund, B., & Berglund, U. (1978). Invariance of odor strength with sniff vigor: An olfactory analogue to size constancy. *Journal of Experimental Psychology: Human Perception and Performance, 4*, 144–152.

Tempere, S., Cuzange, E., Bougeant, J. C., de Revel, G., & Sicard, G. (2012). Explicit sensory training improves the olfactory sensitivity of wine experts. *Chemosensory Perception, 5*, 205–213. https://doi.org/10.1007/s12078-012-9120-1

Tham, W. W. P., Stevenson, R. J., & Miller, L. A. (2009). The functional role of medio dorsal thalamic nucleus in olfaction. *Brain Research Reviews, 62*, 109–126.

Tham, W. W. P., Stevenson, R. J., & Miller, L. A. (2011). The functional role of medio dorsal thalamic nucleus in human olfaction. *Neurocase, 17*, 148–159.

Thompson, R. (1993). Centrencephalic theory, the general learning system, and subcortical dementia. *Annals of the New York Academy of Sciences* 702:197–223.

Tomiczek, C., & Stevenson, R. J. (2009). Olfactory imagery and repetition priming. *Experimental Psychology, 56*, 397–408. https://doi.org/10.1027/1618-3169.56.6.397

Tononi, G. (2004). An information integration theory of consciousness. *BMC Neuroscience, 5*, 42.

Tononi, G., Boly, M., Massimini, M., & Koch, C. (2016). Integrated information theory—From consciousness to its physical substrate. *Nature Reviews Neuroscience, 17*, 450–461.

Tononi, G., & Edelman, G. M. (1998). Consciousness and complexity. *Science, 282*(5395), 1846–1851.

Tufo, C., Poopalasundaram, S., Dorrego-Rivas, A., Ford, M. C., Graham, A., & Grubb, M. S. (2022). Development of the mammalian main olfactory bulb. *Development, 149*(3), dev200210. https://doi.org/10.1242/dev.200210

Turin, L. (1996). A spectroscopic mechanism for primary olfactory reception. *Chemical Senses, 21*(6), 773–791.

Turin, L. (2006). *The secret of scent: Adventures in perfume and the science of smell.* HarperCollins.

Turin, L., & Yoshii, F. (2002). Structure–odor relations: A modern perspective. In R. Doty (Ed.), *Handbook of olfaction and gustation.* (pp. 527–562). Marcel Dekker.

Tye, M. (1989). *The metaphysics of mind.* Cambridge University Press.

Tyler, L. K., Stamatakis, E. A., Bright, P., Acres, K., Abdallah, S., Rodd, J. M., & Moss, H. E. (2004). Processing objects at different levels of specificity. *Journal of Cognitive Neuroscience, 16*, 351–362.

Uchida, N., & Mainen, Z. F. (2007). Odor concentration invariance by chemical ratio coding. *Frontiers in Systems Neuroscience, 1*, 3.

Uva, L., & de Curtis, M. (2004). Polysynaptic olfactory pathway to the ipsi- and contralateral entorhinal cortex mediated via the hippocampus. *Neuroscience, 130*, 249–258.

van Gelder, T. (1990). Compositionality: A connectionist variation on a classical theme. *Cognitive Science, 14*, 335–384.

Vanek, N., Sóskuthy, M., & Majid, A. (2020). Consistent verbal labels promote odor category learning. *Cognition, 206*, 104485. https://doi.org/10.1016/j.cognition.2020.104485

Veramendi, M., Herencia, P., & Ares, G. (2013). Perfume odor categorization. *Journal of Sensory Studies, 28*, 76–89.

Vincis, R., Gschwend, O., Bhaukaurally, K., Beroud, J., & Carleton, A. (2012). Dense representation of natural odorants in the mouse olfactory bulb. *Nature Neuroscience, 15*(4), 537–539. https://doi.org/10.1038/nn.3057

von Bekesy, G. (1964). Olfactory analogue to directional hearing. *Journal of Applied Physiology, 19*(3), 369–373.

Wabnegger, A., Schlintl, C., Höfler, C., Gremsl, A., & Schienle, A. (2018). Altered grey matter volume in "super smellers." *Brain Imaging and Behavior, 13*(6), 1726–1732. https://doi.org/10.1007/s11682-018-0008-9

Wedekind, C., Seebeck, T., Bettens, F., & Paepke, A. J. (1995). MHC-dependent mate preferences in humans. *Proceedings: Biological Sciences, 260*(1359), 245–249.

Weiss, T., Snitz, K., Yablonka, A., Khan, R. M., Gafsou, D., Schneidman, E., & Sobel, N. (2012). Perceptual convergence of multi-component mixtures in olfaction implies an olfactory white. *Proceedings of the National Academy of Sciences of the United States of America, 109*(49), 19959–19964. https://doi.org/10.1073/pnas.1208110109

Wijk, R. A., & Cain, W. S. (1994). Odor identification by name and by edibility. *Human Factors, 36*, 182–187.

Wilson, D. A. (1997). Binaral interactions in the rat piriform cortex. *Journal of Neurophysiology, 78*, 160–169.

Wilson, D. A., & Stevenson, R. J. (2006). *Learning to smell*. Johns Hopkins University Press.

Wilson, K. A. (2021). Individuating the senses of "smell": Orthonasal versus retronasal olfaction. *Synthese, 199*, 4217–4242. https://doi.org/10.1007/s11229-020-02976-7

Wiltrout, C., Dogra, S., & Linster, C. (2003). Configurational and nonconfigurational interactions between odorants in binary mixtures. *Behavioral Neuroscience, 117*(2), 236–245.

Witt, M., & Hummel, T. (2006). Vomeronasal versus olfactory epithelium: Is there a cellular basis for human vomeronasal perception? *International Review of Cytology, 248*, 209–259.

Wnuk, E., Laophairoj, R., & Majid, A. (2020). Smell terms are not rara: A semantic investigation of odor vocabulary in Thai. *Linguistics, 58*(4), 937–966. https://doi.org/10.1515/ling-2020-0009

Woolsey, C. N., & Walzl, E. M. (1942). Topical projection of nerve fibers from local regions of the cochlea to the cerebral cortex of the cat. *Bulletin of the Johns Hopkins Hospital, 71*, 315–344.

Wu, W. (2018). The neuroscience of consciousness. In E. N. Zalta (Ed.), *The Stanford encyclopedia of philosophy* (Winter 2018 Edition). https://plato.stanford.edu/archives/win2018/entries/consciousness-neuroscience/

Wu, Y., Chen, K., Ye, Y., Zhang, T., & Zhou, W. (2020). Humans navigate with stereo olfaction. *Proceedings of the National Academy of Sciences of the United States of America, 117*, 16065–16071. https://doi.org/10.1073/pnas.2004642117

Wysocki, C. J., Dorries, K. M., & Beauchamp, G. K. (1989). Ability to perceive androstenone can be acquired by ostensibly anosmic people. *Proceedings of the National Academy of Sciences of the United States of America, 86*, 7976–7978. https://doi.org/10.1073/pnas.86.20.7976

Yaksi, E., von Saint Paul, F., Niessing, J., Bundschuh, S. T., & Friedrich, R. W. (2009). Transformation of odor representations in target areas of the olfactory bulb. *Nature Neuroscience, 12*(4), 474–482.

Yamazaki, K., Yamaguchi, M., Baranoski, L., Bard, J., Boyse, E. A., & Thomas, L. (1979). Recognition among mice: Evidence from the use of a Y-maze differentially scented by congenic mice of different major histocompatibility types. *Journal of Experimental Medicine, 150*, 755–760.

Yamazaki, K., Yamaguchi, M., Boyse, E. A., & Thomas, L. (1980). The major histocompatibility complex as a source of odors imparting individuality among mice. In D. Müller-Schwartze & R. M. Silverstein (Eds.), *Chemical signals: Vertebrates and aquatic invertebrates* (pp. 267–273). Plenum.

Yao, F., Ye, Y., & Zhou, W. (2020). Nasal airflow engages central olfactory processing and shapes olfactory percepts. *Proceedings of the Royal Society B: Biological Sciences, 287*, 20201772. https://doi.org/10.1098/rspb.2020.1772

Yeshurun, Y., & Sobel, N. (2010). An odor is not worth a thousand words. *Annual Review of Psychology, 61*, 219–241.

Yoshida, I., and Mori, K. (2007). Odorant Category Profile Selectivity of Olfactory Cortex Neurons, Journal of Neuroscience, 27 (34) 9105–9114; DOI: 10.1523/JNEUROSCI.2720-07.2007

Young, B.D. (2011). Olfaction: Smelling the Content of Consciousness. PhD Dissertation, CUNY The Graduate Center.

Young, B. D. (2012). Stinking consciousness. *Journal of Consciousness Studies, 19*(3–4), 223–243.

Young, B. D. (2014). Smelling phenomenal. *Frontiers in Psychology, 5*, 713.

Young, B. D. (2015). Formative non-conceptual content. *Journal of Consciousness Studies, 22*, 201–214.

Young, B. D. (2016). Smelling matter. *Philosophical Psychology, 29*(4), 520–534. https://doi.org/10.1080/09515089.2015.1126814

Young, B. D. (2017). Enactivism's last breaths. In M. Curado & S. Gouveia (Eds.), *Contemporary perspective in the philosophy of mind.*) pp. 96–117). Cambridge Scholars Press.

Young, B. D. (2019a). Smelling molecular structure. In D. Shottenkirk, S. Gouveia, & J. Curado (Eds.), *Perception, cognition, and aesthetics*. (pp. 64–84). Routledge.

Young, B. D. (2019b). The many problems of distal olfactory perception. In T. Cheng, O. Deroy, & C. Spence (Eds.), *Spatial senses: Philosophy of perception in an age of science*. (pp. 148–169).Routledge.

Young, B. D. (2019c). Olfactory imagery—Is exactly what it smells like. *Philosophical Studies, 177*, 3303–3327. https://doi.org/10.1007/s11098-019-01371-4

Young, B. D. (2019d). Smell's puzzling discrepancy. *Mind and Language, 35*, 90–114. https://doi.org/10.1111/mila.12233

Young, B. D. (2020). Perceiving smellscapes. *Pacific Philosophical Quarterly, 101*, 203–223. https://doi.org/10.1111/papq.12309

Young, B. D. (2023). Maybe we don't smell molecular structure. In A. Keller & B. D. Young (Eds.), *Theoretical perspective on smell*. (pp. 168–187). Routledge.

Young, B. D. (2023). Smelling odors and tasting flavors. In A. Mroczko-Wąsowicz & R. Grush (Eds.), *Sensory individuals*. (pp. 243–257). Oxford University Press.

Young, B. D., Escalon, J., & Mathew, D. (2020). Odors: From chemical structures to gaseous plumes. *Neuroscience and Biobehavioral Reviews, 111*, 19–29. https://doi.org/10.1016/j.neubiorev.2020.01.009

Young, B. D., Keller, A., & Rosenthal, D. (2014). Quality-space theory in olfaction. *Frontiers in Psychology, 5*, 1. https://doi.org/10.3389/fpsyg.2014.00001

Young, B. D., & Nanay, B. (2021). Olfactory amodal completion. *Pacific Philosophical Quarterly, 103*(2), 372–388. https://doi.org/10.1111/papq.12357

Zang, Y., Chen, B., & Hummel, T. (2020). Assessment of odor perception related to stimulation modes in a mock MRI scanner. *Journal of Neuroscience Methods, 341*, 108754. https://doi.org/10.1016/j.jneumeth.2020.108754

Zatorre, R. J., & Jones-Gotman, M. (1991). Human olfactory discrimination after unilateral frontal or temporal lobectomy. *Brain, 114*, 71–84.

Zelano, C., Montag, J., Khan, R., & Sobel, N. (2009). A specialized odor memory buffer in primary olfactory cortex. *PLoS One, 4*(3), e4965.

Zhou, W., & Chen, D. (2009). Binaral rivalry between the nostrils and in the cortex. *Biology, 19*(18), 1561–1565.

Zhou, Y., Smith, B. H., & Sharpee, T. O. (2018). Hyperbolic geometry of the olfactory space. *Science Advances, 4*(8), eaaq1458. https://doi.org/10.1126/sciadv.aaq1458

Zobel, S., Hummel, T., Igner, J., Finkelmeyer, A., Habel, U., Timmann, D., Schulz, J. B., & Kronenbuerger, M. (2010). Involvement of the human ventrolateral thalamus in olfaction. *Journal of Neurology, 257*, 2037–2043.

Index